Sugar off !

Richard Cook qualified  in
hospital and dental prac
Anglia. He now runs a fi
The Times Educational S
history and literature, an

Elizabeth Cook worke
television music program ....-year spell in
Africa.

Richard and Elizabeth Cook have both had a long-standing
interest in food, diet and health, which has become more sharply
focused since the birth of their first child. They now have four
children, and are active campaigners for healthier food and
eating patterns.

Richard and Elizabeth Cook

Sugar Off !
A practical guide to sugar-free living

Pan Books   London and Sydney

First published 1983 by Great Ouse Press, Cambridge
This revised edition first published 1985 by Pan Books Ltd,
Cavaye Place, London SW10 9PG
9 8 7 6 5 4 3 2 1
© Richard and Elizabeth Cook 1983, 1985
Illustrated by Duncan McLean
ISBN 0 330 28633 1
Photoset by Parker Typesetting Service, Leicester
Printed and bound in Great Britain by
Hazell Watson & Viney Limited,
Member of the BPCC Group,
Aylesbury, Bucks

Dedication

For all children and especially our four arch-food-testers, Marjorie, Alice, Edward and Eleanor in the hope that all they will ever see of a dentist's work is a mirror and probe.

Contents

Preface to the Pan edition

Three days after the publication of the first edition of *Sugar Off!*, *The Sunday Times* carried a front page article by Geoffrey Cannon highlighting the failure of the DHSS to publish a report on diet and health which it had commissioned. This had been prepared by a working party of the National Advisory Council for Nutrition Education (NACNE). It argued that the ordinary British diet is a major cause of ill-health and premature death, recommending that the intake of fat, sugar and salt be reduced during the '80s by 10%, while that of fibre be raised by 25%.

The blocking of the report under pressure from the food industry was probably counter-productive in that it provided a focus for the widespread concern many people have about some aspects of food policy. We felt that such concern ought to be reflected in the new edition, by bringing our advice and recipes more closely in line with NACNE.

Sugar Off! of course was well ahead, not only with the short-term aims of the report but with its longer view that sugar consumption be eventually reduced by 50%. Our use of dried fruits and whole grain cereals took care also of the fibre requirement, while we had always been sparing with salt. However, although general warnings about fat, and especially animal fat, occurred throughout the book, we felt that we should now be more specific and positive. In general we have now replaced hard fat ingredients by vegetable oils, especially soya and sunflower which are high in polyunsaturates. We have also reduced actual oil:cereal ratios in the pastries, and where possible have come down to scone mix levels. Whenever baking powder is used to lighten texture we have removed the salt ingredient to keep overall sodium levels low. Although NACNE had been cautious on dietary cholesterol and declined to make recommendations, we have played safe here also. Heart specialists suggest that we eat no more than half an egg a day, and we have trimmed accordingly.

A positive outcome of these adjustments was that they suggested new fields of investigation and this edition has, overall, more recipes than the first.

Our coincidence with NACNE and the need to link dental health more closely with general health took us into another interesting area. In planning the book, our first task had been to discover where sugar occurred in foods, and we wrote to every expert we had heard of for help. We drew a complete blank. It seems that as late as 1975 nobody had been interested enough in the sucrose content of foods to estimate it. Sucrose, other sugars and starches were lumped together under the 'carbohydrate' label, and we had long been taught that these were the nutritional baddies. That potatoes and bread made us fat was part of the same conventional wisdom which included a horror of skimmed milk as unfit for babies, and the certainty that dairy foods were essential to the diet.

That this wisdom has now been substantially revised has created a credibility problem which seems to have gone unnoticed by nutritionists and dieticians. People have not been slow to wonder whether nutritionists, having been wrong once, might well be wrong again. It seems that all professions share a prim unwillingness to admit past error, which blights the impact of their revisions. As it happens the last two decades have seen an explosion of scientific effort which is now replacing previous guesswork, and we believe that the nutritionists are now getting it right. But all of us would do well to take the heart the sound common sense of Sir Peter Medawar:

> Today's opinions may not be the same as yesterday's, because they are based upon fuller or better evidence. We should quite often have occasion to say 'I used to think that once, but I have now come to hold a rather different opinion'. People who never say as much are either ineffectual or dangerous.[1]

Good science deserves better marketing than it gets, especially now that it is replacing the shallow thinking of the past.

For our part we hope that we are making the dental health message more clear and that our perfect teeth will be housed in long-life bodies refreshed by healthy arteries and bowels.

Preface

The idea for this book was born at the same time as our first child in 1972. Although we had lived cheerfully for years on the proceeds of routine dentistry practised upon other people, the prospect of visiting the same on our own child filled us with horror. How could we protect her from this awfulness?

We looked first at what traditional dentistry had to offer and found little but imprecise exhortations connected with dentists and bathrooms. Its track record, we knew, was not impressive. Certainly traditional practice could claim little relevance to the problem of actually *preventing the onset* of dental decay.

The newer style, known as 'preventive dentistry', seemed at first more promising, and we signed up enthusiastically for seminars where it was evangelized. From them we made the valuable discovery that *positive* dental health (i.e. not actually having teeth go bad), was a subject quite separate from the dentist's normal work and possibly even unconnected with it. However the burden of such seminars was to show how prevention could (and should) be corralled within the limits of dental practice. It seemed that only those matters susceptible to professional control: fluoride applications, fissure sealings and intense oral hygiene instruction, were to form the scaffolding of the new dentistry. Moreover a strong feature of the new movement was the huge increase in job satisfaction which was promised on conversion. We even read statements to the effect that no dentist who had taken the preventive path had ever come to regret it. Later we were to meet some who had, but at the time we felt entitled to wonder at the nature of a preventive dentistry that left the dentists doing very nicely for themselves as well as increasing their job satisfaction.

It finally dawned on us that however enthusiastic and forward-looking such dentists were, they were in fact kept in business by a powerful sugar culture which could bludgeon through any orchestration of painting, coating and scrubbing of teeth. Any suggestion in preventive circles, however, that this culture ought to be a significant, if not the main, target for our energies was almost invariably met with patronizing smiles.

We found, however, that among patients and other people, particularly anxious parents like ourselves, the idea of finding a healthy path through our sugar-drenched culture was very welcome. We decided therefore that dental health was too valuable a property to be left in the exclusive control of dentists and decided to go it alone with this book. It is the fruit of seven or eight years' investigation, not in dental surgeries or scientific laboratories (though we do use their results), but at source, in kitchens, factories, shops and supermarkets, schools and playgroups, where the significant action takes place, and where we received much positive and practical encouragement.

Although, in general, dental colleagues viewed our efforts on a scale calibrated between the knowing smile and blank uninterest, we would like to record our gratitude to the shining exceptions. First we thank Colin Dexter, at whose seminars the idea of an actual book germinated and began to seem feasible. We owe a special debt to Dr Aubrey Sheiham of the London Hospital who placed his scholarship at our constant disposal and guided our grasp of dental sciences from its rickety beginnings to its present modest level. We thank Professor W. M. Edgar of Liverpool University for patience and thoroughness in dealing with our queries. It should be unnecessary to add that none of them are to be thought responsible for any views expressed in the book, or its overall philosophy.

During the last three years we have received constant support on a broad front from Michael Craft, Ray Croucher and their team at Cambridge University's Dental Health Study. At several of their events we have field-tested the recipes and had opportunities to discuss them with interested parties, gaining useful feedback. With their help we were able to test two of the puddings at Caverstede Nursery School in Peterborough. Our special gratitude in this instance goes to Mrs Jackson the cook, head teacher Betty Hay and her staff, who kept faith with us in spite of a rice pudding disaster for which we now make public apology to the children concerned. The recipe has now recovered and we hope they have as well.

Hardly any recipe book was of help in this new venture and we looked to sympathetic friends for help. A major source of encouragement came from our old friend Hazel Pope who sparked off many of our new culinary thoughts in general, and who in particular offered guidelines for the nut butters. To Gretchel Croucher a bouquet also for leading us to a barbecue sauce.

As novices in the field of food technology we had not merely to read the standard textbooks, but also to consult the nutritionists and food chemists of many different firms directly. We want to thank all of these, alas too numerous to name, for their kindness in seeking out relevant information even when the products with which they were involved were to be victims of our special interest. Many went further and offered their good wishes for the success of our enterprise and we would like them to know how much they cheered us.

We thank Professor John Yudkin for agreeing to read a draft manuscript and for his many useful comments and suggested corrections. Among friends who read the script we thank Dilys and Edgar Page, Dr. J. D. A. Miller, Dr John Besford, Kate Cook, and Bill McKeith, whose critical observations helped to give us an outside perspective when we were perhaps over-immersed. Finally we remember two old friends Donald Brook and David Mercer, who, a few weeks before the latter's death, lifted our flagging spirits by laughing like drains at our final draft.

Richard and Elizabeth Cook

1 Rot, rule and ritual
The tooth advice muddle

As if Religion were intended
For nothing else but to be mended.

Samuel Butler

Once I had a horrible toothache. I was getting on in years
and one of the semi-criminals, a kindly good natured youth
who suffered the pain as if it was his own, whose own teeth
whether from youth or some more mystic cause were in
really good nick, marvellous condition, said to me with deep
concern, 'Excuse me, Andrei Donatovich but you must be a
terrible sinner. Your teeth are always aching whereas mine
for instance never have yet . . .' I couldn't argue. Evidently
that was the way of things . . .

A. Sinyavsky

Everybody agrees that prevention is better than cure. Certainly it's
the standard thing to say even when we are not quite sure what we
mean or don't feel inclined to do much about it.

But is prevention always as possible as cure, however desirable?
We all suspect that if we ate more sensibly, cut down on the
boooze, stopped smoking and took more exercise we would certainly be healthier. There remains, however, a number of diseases
that all the precautions in the world can do little to forestall.

Dental disease isn't one of them. Nobody need ever have bad teeth
or bleeding gums. Enough is now known about the causes for us to
be able to prevent them completely. So why is this marvellous
prospect not on conspicuous display? How is it that the dentists'
waiting rooms are still packed with sufferers from these totally
preventable conditions? Either the means and information are not
being put about, or if they are they are not being accepted. The

first thing to look at then is the nature of what is actually made available.

There is in fact no end of advice about how to keep your teeth. Health Clinics dish it out, toothpaste advertisers feature it, and dentists' waiting rooms may be plastered with it. Wherever it occurs it seems to shake down to three golden rules; brush (usually gums to teeth) twice a day or after meals, visit the dentist twice a year, and watch the sweet stuff, particularly between meals.

Somehow it hasn't worked. The statistics[1,2,3] show that, for all the money spent on dentistry and toothpaste, British teeth, at least, inhabit a disaster area, and other evidence is all around. People who have attended the dentist routinely, watched the sweets and brushed regularly still end up in dental cripplehood. Dentists complain of overwork. We all know people whose breath could strip paint at twenty paces. Ruined gappy mouths haunt us from all sides. Practically everyone over fifty has the same face. Every day that fearful symmetry below the nose offers its xylophone grin from the bingo halls and studio audiences. So perhaps these rules need to be put under closer scrutiny, to investigate their relevance.

Brushing and the pastemakers

In the days when not much was known about genuine prevention and tooth-rot seemed inevitable, much attention was paid to toothpastes, or, to give them their genteel title, dentifrices. It was thought that since a lot of the stuff was being sloshed around a lot of mouths some magic ingredient could be incorporated that might kill the germs, make the teeth superclean, neutralize the acids or absorb the bad smells. Remember chlorophyll, GL 99 and all the other wonders that gave our mouths a temporary lift but never quite did the trick! The magic, alas, stayed with the pretty girls in the ad and it took a long time to get into the tube.

When it finally did get there, in the form of fluoride, all the previous claims were dropped. No longer were we to be delivered from social ostracism and assured of sexual success by paste use. What we, or at least our children, were now offered were 'fewer fillings'. The pastemakers had finally got round to peddling health.

This change of emphasis is welcome and some of the improvement in children's dental health now being recorded may well be related to it. Little else has changed. In particular the 'how' and 'when' to brush still remained as pillars of certainty in a changing world. We were to brush from gums to teeth and we were to do it *after* meals. No matter that all the studies[4,5,6,7,8] show that, of all the methods, this one is probably the least effective, we learned like Orwellian sheep to know, 'up and down good, side to side bad'. The method gets universal applause and the rational dentist who is brave enough to counsel otherwise finds her or his efforts sabotaged because the day before some TV personality, presumably with professional advice, has been asserting the goodness of up-and-downness. Meanwhile the glum statistics inform us that when the fluoride element is subtracted from the scene, the brushing that most of us have been doing has made little difference to our dental health.

'Up and down good, side to side bad!'

In the first edition of Professor G. N. Jenkins's standard work, *The Physiology of the Mouth* (1956) the author suggests that it makes much more sense to clean teeth *before* meals in order to remove the bacterial plaque which ferments the dietary sugars. In the mountain of dental health education material we have examined, only one work advises this.[9] All the others have us racing to the washbasin at the drop of a spoon.

That other fixed star in the preventive firmament is the six-monthly dental inspection. Nobody knows how this one started, but an American dentist[10] who researched it could find no earlier reference than the 'Amos and Andy Show', sponsored by the Ipana Tooth Paste Co. in the 1930s. The slogan of this radio show was 'Brush with Ipana twice a day, visit your dentist twice a year'.

Many dentists appear to think that the instructions came down from Sinai with the Tablets. There is no reference, however, to show that scientific research had any say in the matter. Indeed a study[11] published in 1977 could find no scientific basis for it. A survey[12] of children in 1973 showed that regular attenders don't have much advantage over the irregulars. They have more fillings but no less disease. A Swedish experiment[13] in 1978 demonstrated that the routine dentistry proceeding from regular dental inspections is 'highly ineffective' in preventing disease. In fact the most hopeful analysis of statistics[14] we could find showed that if you are going to lose all your teeth, you might by dental attendance delay the event by up to five years.

Sweet, sweet poison for the age's tooth

We all know that this is where it starts, that it is our profound addiction to the sugary stuff that produced the problem in the first place. Advice about it is, however, perfunctory. For instance, a glossy handout to dentists from the makers of Oral-B toothbrushes, 'Caring for your baby's teeth', explicitly avoids dietary advice. 'All foods and liquids you feed to your baby probably contain acid-forming sugars that contribute to this dental decay. So it's never too early to start practising good oral hygiene.' Sugar gets no further mention.

It is easy to see why. Dietary control will not sell a single tube of toothpaste or earn a penny for the dentist. So they say, 'Watch the sugar, especially between meals.' It is the one commandment with a fair measure of scientific backing.[15]

Unfortunately, dentists are trained to mend and pull teeth, not to advise on dietetics, and anyone who has ever watched them at their gatherings can be in no doubt how they view the importance of diet. Writing in the *British Dental Journal*, 3 May 1977, Elizabeth Elliott, (Lay) Executive Manager of the British Dental Health Foundation, asked,

> . . . why it is that so many of the profession who are
> preventively orientated are so frequently seen eating the biscuits
> richly embellished with jam, chocolate and sugar that are

provided at virtually all meetings organized by dental societies, hospitals and members of the dental trade. At a recent seminar held in a hotel which included a lunch it was only with great difficulty that a colleague and I were able to obtain cheese and biscuits instead of the pudding that was mostly sugar.

The letter went without answer.

In one specialist children's practice we have visited, sugary cordials were available free in the waiting room, and in another it is routine to reward 'good' behaviour in the chair with a sweet. It is also routine dental practice to recommend syrupy teething remedies and in the NHS Dentists National Formulary there are no fewer than 15 items in which the drug is delivered in syrupy form (i.e. 66.6% sugar solution).

Human rites

The trio of injunctions then is not only token. It has very little basis in science. It belongs to a time when nothing much was known, and nothing much could be done, about dental disease, and ignorance always looks less frightening when tidied up a little. The traditional role of the dentist has always been to relieve pain, repair the damage and restore the gaps. If anyone talked about dental health it would be in connection with these three procedures. The rules were a sort of ritual incantation against the dark, and dentistry a kind of sub-religion administered by much-feared priests. For original sin we had 'soft teeth' and 'weak gums'. Guilt would drive us to a daily ritual with a toothpaste of assured talismanic powers, and to token self-denials like 'one spoon, not two'. For the very young, an initiation ceremony – He's two, three or whatever. When shall I start bringing him? I want him *to get used to it*. For the older ones a six-monthly presentation to the priest who might order penances in the form of painful procedures or the highly artistic ritual mutilations of restorative dentistry, while for the deeply errant a bloody sacrifice might ensue. For the prodigal non-attenders, the comfort of confessional, 'I've been naughty. I haven't been for two/three years. I'm ready to take what's coming. *Punish me!*'

The rules were a catechism, important not for their accuracy but for their rigidity that gave the people certainty and protected them from alarm and confusion.

But the age of simple faith is over and the relationship of priest and believer, dentist and patient, has to be re-negotiated. Science now makes it possible for us to maintain our own dental health. In fact, as this book will show, such steps have little necessary connection with dentists, though a growing number of them have adapted their practising styles and *can* offer help to the DIY patient. But no dentist can bestow or dispense dental health. This you have to get for yourself. There are many obstacles and no magic formulae. This book is about those obstacles and how to get round them. It is a new and hitherto uncharted area. Those who travel through it will gain in self-reliance. Others who prefer to leave it to the professionals still have the dentist with whom they can dock their mouths twice a year. What they will get there, however, are those depressing three Rs: relief, repair and replacement. For those who want something better there remain the following chapters and in the first of these we are going to settle for good the question of title.

2 Whose mouth?
The facts of tooth rot

Such a day it is when time
piles up the hills like pumpkins,
and the streams run golden.

When all men smell good,
and the cheeks of girls
are as baked bread to the mouth.

As bread and beanflowers
the touch of their lips,
and their white teeth sweeter than cucumbers.

Laurie Lee

English Teeth! HEROES' Teeth!
Hear them click! and clack!
Let's sing a song of praise to them –
Three Cheers for the Brown Grey and Black.

Spike Milligan

For the new-born baby the mouth is the focus of all pleasure. Through it pours the warm milk that soothes an empty and aching stomach. Lip, cheek and toothpad are fulfilled by contact with the warm breast. All new objects are passed to this arbiter for inspection. Pleasure is registered by a stretch of lip-rim into a wide joyful circle. As affections develop, tongue, lips and cheek are involved in their display.

In the beginning all is sweet and delicate. The smell of a baby's breath is curiously compelling. Give or take an occasional throat infection, this is how it stays. With the arrival of teeth, however, the smells gradually modulate. This pink innocence can become within months, and certainly years, a centre of corruption. Smells there are, but avoided. Mouth-to-mouth contact becomes a qualified pleasure. By adulthood it may be removed from the repertoire entirely.

The reason for this decline and fall lies in the mouth's form and role. It is an apparent break in the body's defences through which bacteria have easy access. It is a wet, warm place, an ideal environment for their growth. In general, the dominating bacteria are harmless and, by crowding out their more dangerous fellows, give a high degree of protection.

There is, however, one exception to this overall benignity and from it proceed the smell, the decay, in fact the whole absurd and unnecessary world of dentistry. The villains are a group of bacteria which have the capacity to tack themselves on to gums and teeth by means of a starchy glue which is their food supply. The combination of bacteria and glue is now known as 'plaque', a name which the paste manufacturers have tattooed on to the public consciousness. The plaque at the junction between tooth and gum can be particularly destructive. It is, however, only destructive to teeth when fuelled by a flow of high energy sugars.

In contemporary mouths, alas, this flow is seldom stilled. Day by day and often hour by hour, the sticky growth is nourished and swelled by the constant presence of sugar. In its depth there is a constant manufacture of the acids which soften and rot tooth enamel, the hardest substance in the human body. Year by year victims are claimed and the stench of corruption increases as the mouth collapses into ruin. The removers are called in. The force-fed plaque has been all too successful.

It is the supervision of this long death which sustains the vast profession of dentistry, a profession of rituals confined within an overall expectation of ultimate failure. The final act of this supervision is the installation of a monument to defeat, a non-biodegradable grin which will outlast its owner's dust and ashes. In the mouth its plastic gum margin, while not as effective as flesh and enamel, can still, however, supply a launching pad from which the sugar-made stenches can be pumped on to the breath.

It is at this point that we must bring our theology up to date. The incrimination of plaque as the newly-revealed fount of evil has brought a new sacrament called 'plaque-control', administered by a new priesthood of so-called 'preventive dentists'. From the

24

United States, a steady stream of smartly-suited evangelists has poured into Europe to teach the new rites. At the same time the pastemen, adapting themselves hurriedly to the changed order, have gone into full production of the indispensable aids necessary for the new observances. Serious dental scientists in their employ have devised control experiments to prove the superiority of one brush over another as a plaque-remover, and their results have been solemnly reported in the advertisements. It has become a world of gadgetry and the message has gone out that plaque is the single obstacle preventing dental health and plaque-control the golden key to a glowing future.

But before we leap too precipitously into this brave new world, with our tots moving into plaque control as into the first battles with Lego, we should perhaps examine this novel pestilence a little more closely. Can it for instance be the mouth's norm to be subjected to mechanical sterilization on a daily basis? Is this obsession with cleanliness to be a measure of health? Should we not remember that plaque was always present on human teeth and that only in recent years has it begun to rot them?

Examinations of the skulls of ordinary people who lived in the Middle Ages and before,[1,2,3,4] show that decay then was rare. When it did occur it was first of all, as now, in the fissures of young back teeth. By the late teens the ordinary diet of that time had wiped out both fissures and decay, the latter not re-emerging until late in life, when heavy wear had broken through contact points between adjacent teeth to permit food stagnation. Amounts of plaque were high but amounts of decay were small. Decay, as we now know it, only occurred in the mouths of the rich who could afford sugar priced like present-day caviare.[5] This was the fuel that turned the Jekyll plaque into the modern and murderous Hyde.

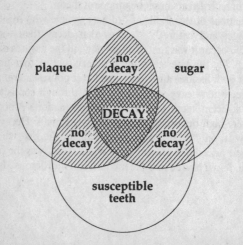

Preventive dentistry then, and its obsession with plaque, represent the response of both dental profession and pastemakers to the threat of losing control of the situation. The interaction of the three decay-producing factors shown above shows how serious that threat is. Little of what is displayed has any necessary connection with dentists, though much is made of their role. The circle representing sugar tends to be shrugged off by them as a hopeless cause, something at the mercy of mere patients who can conveniently be blamed when the prevention package fails to deliver. In fact the equivalence of the three circles is in any historical context an exaggeration. In terms of importance there is one great circle representing the world of food to which we apply two minimal circlets within which the dentist can play a minor role. In truth dental surgeries hardly come into it. The battle is located and fought in the kitchen and in the supermarket.

So we again have a trio, albeit an unbalanced one, this time not of commandments, but of targets for our self-helping energies. We are now into Tooth Liberation and we know whose mouth we are talking about. It is neither a dentist's adventure playground, a pastemaker's washboard or a sugar-pusher's rubbish bin. It is and will be a pleasant piece of private property to have and foster. That baby's sweet-smelling pinkness can persist free of disease and mutilation. The breath that moves across it need never embarrass the breather nor cause a lover to flinch. With this assurance we can now move into the obstacles and clear a way for self-service dental health.

3 A kitchen disease
The sugar invaders

. . . ladies' lips, who straight on kisses dream;
which oft the angry Mab with blisters plagues,
Because their breaths with sweetmeats tainted are.

Shakespeare
Romeo and Juliet I IV

. . . a hooked nose, narrow lips and black teeth, a defect the
English seem prone to from their too great use of sugar.

Paul Hentzner
describing Elizabeth I

There ought to be two ways in which the food we eat can affect our
teeth. First, since we are composed of what we have eaten, this
composition might find reflection in tooth quality. Hence most
dietary advice on dental health is padded out with routine stuff
about getting the right balance of protein, fat, carbohydrate, vitamins and minerals. It is often accompanied by drawings of slabs of
cheese, bottles of milk, carrots and apples, to help us identify them
in the shops. There isn't, however, much in the way of hard
evidence that links failure to get this balance with bad tooth
structure or chemistry.

While it is known that certain vitamins, together with minerals
such as phosphate and calcium, are all necessary for healthy bones
and teeth, it is very rare, at least in developed countries, for these
to be lacking in a child's diet. Moreover, outside of folklore, lack
of calcium or vitamin deficiencies are rarely connected to tooth or
gum disease. The fact is that nobody knows how to *make* good
teeth out of meals. What *is* known is that a commercially inspired
'vitamania' can certainly make teeth go bad. This phenomenon
will be examined in the next chapter, but here we have to

emphasize that the effect of food on teeth is predominantly local. It is what happens as a result of food passing through the mouth that should claim our attention. It is the food which fuels the plaque and keeps its acidity high that is the centre of concern and, without doubt, the major criminal in this respect is sugar.

There is a whole family of sugars, most of them produced by plants. The sugar most involved has the chemical name of *sucrose* and this is the stuff we buy in bags. Whether it appears in lumps, grains or powders, white or brown, or even those insoluble coloured crystals you get in genteel restaurants with your thimble of coffee, it is all sucrose. Other sugars, like *glucose*, *fructose*, and *maltose*, are just as fermentable by the plaque but they are more expensive and less commonly used. Moreover the plaque bacteria responsible for decay seem to prefer the larger sucrose molecule and from it obtain the energy to weave the starchy glues that clamp acidic end-products on to the tooth surface.

The mechanism of sugar's involvement has been thoroughly examined. Experiments have been done to show that when a sugar solution is dripped on to plaque-covered teeth the acidity against the tooth surface rises to danger point within a minute or so and stays that way for another twenty minutes.[1] If the drip is continued at two-minute intervals this acidity remains at danger level, i.e. at a point where enamel can be dissolved. Obviously then, frequency of sugar presence is the important factor rather than sheer quantity. That it works that way in practice was shown by a famous, and rather immoral, experiment in Sweden,[2] when a study was made of 400 mentally defective patients. These were split into groups. One group had no sugar, another only at meals, while a third was given sugar in the form of toffee to be eaten at will. This last group got the most new cavities, while those who ate sugar only at meals were not much worse off than those who had none.

These two studies are the foundation on which the dental conventional wisdom about diet is based. This is that sugar consumption should be reduced, especially between meals. It is a simple enough conclusion seen from the uncomplicated standpoint of the scientist and professional. Sugar is bad for the teeth. The way to keep teeth healthy is to eat it less frequently. The way to get people to eat less of it is to tell them to eat less. The social scientist might well find this approach breathtakingly naive, while a lay response could be that all this talk about frequency suggests that what is all right at mealtimes can't be all that bad in general. Moreover this obsessive incrimination of 'frequency' enables the sugar-pushers to distract attention from its subject. But of course the instruction is not given in any realistic sense. It is merely part of the catechism. Any serious inroad into the resolution of sugar-borne disease must start not only with sugar as chemical, food, or plaque-nourishment, but as real people actually think about it, i.e. with sugar as magic, medicine, and symbol.

In the beginning there was no sugar in the sense that we use the word, not outside the South Pacific that is. There was honey, of course, and it was valuable enough to act as currency for rents. In Northern Europe and Russia it was traded down the great river roads against the sophisticated exotica of Byzantium. We know also that it was involved in magic and religion. In the early Christian Church honey figured in the rite of baptism, while its pagan connection with purity and virginity continues in the obligatory use of beeswax for votive candles by the Roman Catholic Church. But although the wealthier of Saxon farmers may have made free with it, particularly in the form of mead, much of the sweetness vouchsafed to the pre-medieval palate would have come from fruits. The sun's energy was directed by these Saxon farmers predominantly into the starchy grain and its palatability was easily raised by the *condiment* sweetness levels of the fruit. A rough enough old diet, gritty enough certainly to grind away the plaque-secreting grooves and fissures where much modern decay starts. It looks also as if they got the sugar level just about right.

But the sugar cane, indigenous to the South Pacific, had reached into mainland Asia and spread into Persia by the sixth century AD. From there, in the period of Arab expansion, it spread through the Mediterranean. By the end of the fifteenth century, it was taken to Madeira and the Cape Verde Islands, reaching the New World in Columbus's second voyage of 1493. It was still a costly item. One of the earliest references to sugar in English history is that in 1176, Henry II 'bought for his own use 34 pounds of sugar at a rate of 9 pence a pound'.[3] Its mention continues to be rare until the fifteenth century when its use was still confined to the rich. At this time there was also a significant drop in honey production. By 1558 'in the Furstenwald only 187 hives remained where once 1,029 had been counted'.[4] It is also about this time that a modern pattern of dental decay begins to show in the mouths of the well-to-do.[5,6]

Gradually, demand for the new sweetener increased. The sweetness that marked the nutritiousness of a food item could now be dissected out and taken neat. The little bit that we fancied became the constant titillator we couldn't do without. It became, if not the main course, then a substantial part of it. The manufacturers quite literally couldn't produce enough of it and, since the colonists of the New World had killed off most of the indigenous populations in the producing areas, they were faced with a labour shortage. It is at this point that the addiction to sweetness gives rise to some very nasty consequences indeed.

While it may have been likely, there is no direct evidence of the systematic use of slaves in sugar production before it began in Western Europe. In 1444 a cargo of Negroes was landed at Lagos in southern Portugal. This was the start of a slave traffic that was to continue for four centuries. The economy of the New World soon became heavily dependent on this traffic. During the four centuries more than twelve million slaves were landed and, if we take into account the huge losses from sickness and sea disasters, the total abduction figure is much higher.

The collection loss cannot be estimated, but it will be no exaggeration to put the tale and toll of the slave trade as 20,000,000 Africans, of which two thirds are to be charged against sugar.[7]

We have therefore a sugar culture with slavery at both ends. At this end, those involved in sugar manufacture are committed to ensuring that this enslavement shall not be easily redeemed. To see how much the new slavers have going for them, we have to go back to the foundations of the sugar culture in which we partake, as with honey, of myth and legend.

For instance the Polynesian Adam and Eve story is of a fisherman who discovered the first woman enfolded in a sugar cane. In Indian mythology, Ikshvaku, son of Manu the father of all mankind, was found as a child in a sugar cane field (the Sanskrit word for the cane is *ikshu*). This was the foundation of a sugar cane dynasty which in the hundredth generation gave rise to Gautama, the Buddha.

There is even an Indian Cupid, Kamdeva by name, whose bow is of sugar-cane and whose string is made of bees and honey.

He bends the luscious bow and twists the string,
With bees so sweet, but ah, how keen the sting!
He with fine flowers tips the ruthless darts
Which through five senses pierce enraptured hearts.

In whatever form, sugar or honey, the concentrate must have figured as a distillation of the sun's power, as crystallized sunlight to carry a kind of warmth through the long uncharitable winter months. And if in these days of confident supply and the banalities of the supermarket the magical aura seems to belong merely to history we should perhaps look again.

He claimed to know of the existence of a mysterious country called Sugarcandy Mountain, to which all animals went when they died. It was situated somewhere in the sky, a little distance beyond the clouds, Moses said. In Sugarcandy Mountain it was

Sunday seven days a week, clover was in season all the year round, and lump sugar and linseed cake grew on the hedges.[9]

The novelist's metaphor for an animal paradise derives naturally from the equation our language makes with sweetness and sugar, on the one hand, and desirability and happiness, on the other. Thus an elderly lover is a sugar-daddy and love is the sweetest thing. Lovers are honeys, sweethearts and sugars. But sugar is above all, big money and much of this is spent in shoring up the good image the sugar culture has bestowed for free.

A popular error, which it is desirable to refute, is the alleged injurious effect of sugar on the teeth . . . Sugar and sugary sweets promote a copious flow of saliva in which they are quickly dissolved, and the solution is swallowed so rapidly that it is removed from the teeth before bacteria can commence activities. The soft pulpy foods such as meat, bread and vegetables, whose texture and insoluble nature cause their retention in the teeth, favour the growth of bacteria and pave the way for dental decay.[10]

The writer was himself a sugar manufacturer and our comment must be of the 'Well he would, wouldn't he?' sort.

However, that sort of crude selectivity has recently given way to a much slicker and more sophisticated approach. In 1977 the British Sugar Bureau circulated the medical (but not the dental) profession with its sleek *Questions and answers on sugar* (now (1983) re-issued more or less unchanged as *Sweet Reason*).

Students of the glossier forms of casuistry and in need of a laugh could do worse than apply for a copy to the British Sugar Bureau, 140 Park Lane, London W1Y 3AA. When they finally get round to admitting the role of sugar as prime plaque food it is as a prelude to a celebration of the paramountcy of oral hygiene in decay prevention. No scientific evidence is given for this assertion and we still await an answer to our 1977 request for this.

But such clumsy public relations exercises represent only the small arms of the sugar-pushers. The really big money goes where the magic is – in advertising. Here contemporary sorcerers have absorbed carefully what the sugar culture gives them for nothing, to distil a modern myth. The notion of 'crystallized sunlight' in this understanding leads naturally to a linguistic legerdemain in which energy, in the sense of 'zip' or 'zest', is available cheaply by the spoonful, the lozenge, or mouth-sized snack. Millions of pounds are spent to persuade us that as we flag between meals, or feel a little off-colour, a sugary snack will supercharge the engine, when of course all it does is to fill up the tank, which seldom empties in any case. Against the common sense view that washed-out feelings are more often related to smoking, booze, over-work, lack of sleep or *over*-eating, the admen offer a primitive but power-ful magic, A Sun God in bags, to blaze through our fatigue and energize our flagging bodies. By a quite trivial expenditure we can be guaranteed sexual and social success, health and zip. And if we make the miserable discovery that all the toffee money in the world can still leave us unloved, impotent, unpopular and fat, and mount a mighty intellectual effort to be rational about what we eat, there is an ambush already prepared.

Even the most dietetically illiterate of us knows that there is a lot of sugar in toffee, cakes and puddings. But do we expect it in corned beef, luncheon meat, frozen chips, cholesterol-free mar-garine, curry mixes and canned vegetables? We may not but we can get it. By law most canned and packaged goods (except some biscuits) must carry an ingredient list on the label in descending order of concentration. The interesting thing about these lists is not how often sugar occurs in unlikely products, but how far up the list it usually is. Until the sugar crisis of 1974 it used to take first place in Heinz Tomato Soup and even now it has only dropped one rung. It is number 1 in Camp Coffee and Crosse and Blackwell's Branston and Tangy Pickles, coming second in Wait-rose Pease Pudding, Kraft's Coleslaw Dressing and most other tomato soups and piccalillis. Anyone who wants to lose weight by the use of so-called slimming foods had better learn to read the small print on the labels. They are full of the stuff.

Against this heavy artillery we will not go very far with a mere instruction to watch the sugar, especially between meals. We have to devise a simple counter-strategy. This must start from an under-standing of the kinds of food with which our bodies have been evolved to cope. Both the design of our teeth, and the nature and length of our digestive tract suggest that we are suited to an omnivorous diet, i.e. one composed from animal and vegetable sources. Teeth and tract together constitute an extraction plant from which the body draws its nutrients. Increasingly, however, technology is taking over this function. The sugar, oil and flour refineries now ensure that what is presented to the digestive system is already half-digested. Our bodies are now challenged by concentrates which they are ill-equipped to manage and have become victim to a sorry series of food-related diseases: obesity, haemorrhoids, diverticulitis, diabetes, tooth-rot, constipation, and so on.

One reaction to this tendency has been the growth of the wholefood movement to ensure the supply and use of unprocessed foods, which must be seen as the obvious ally to the purposes of this book. It has a great deal to offer. Its products are often tastier than those of the supermarkets and it does not embarrass us with glossy, wasteful and irrelevant packaging. It has filled the gap left by the demise of the old-fashioned dry-goods grocer, stocking an ever increasing range of pulses, grains, nuts and fruit, to enlarge our culinary horizons in the direction of both taste and health. In our experience their goods are invariably fresher than the supermarket equivalent, for all the latter's technology. Moreover, to rejoin our central concerns, there is a genuine link between the wholeness of some cereals and nuts and dental health.

However, even this movement has not been able to escape being taken in by the sugar culture. Consider this notice taken from a wholefood shop display.

THIS CAKE CONTAINS NO SUGAR. IT IS SWEETENED
WITH MOLASSES, HONEY AND CAROB SYRUP ONLY.

Here sugar has been avoided by calling it something else, a practice, incidentally, not unknown to the junk food manufacturers with their dextrose, malto-dextrins, corn syrups and the like. A straight substitution of under-refined syrups without a drastic reduction of sweetness levels takes us nowhere. Such levels were determined historically by the growing cheapness of sugar. Later we will show how sweetness can be achieved without recourse to the concentrates, but meanwhile those who wish to regard syrups, malt extracts, and brown sugar (96% sucrose) as 'whole' should employ approximately a quarter of the amounts used in orthodox recipes. This must have been the level of use of those great bee-keepers and dried fruit importers, the Romans and Normans

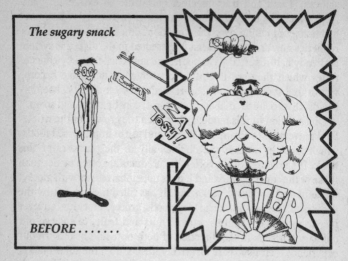

The sugary snack

BEFORE . . . .

who usually kept their teeth for a lifetime with only minimal amounts of tooth decay.

Although if we returned to the Romans' or the Anglo-Saxons' diet we might end up with teeth as healthy and, given a bit of tooth-cleaning, more healthy, such a diet would be unlikely to be a best-seller. Good food there was and Miss Dorothy Hartley has described a twelfth-century way of preparing vegetables,[12] for instance, which could put to shame any contemporary cook, for all our labour-saving equipment. There was also that little matter of seasonal variation in supply, which might well have left our ancestors with scurvy and poorly-fleshed bone come the spring. There is much we can learn from them, however. A diet composed of milk,[13] whey, cheese, bacon, maslin bread and pease seems to have been the staple of the fifteenth-century peasant. To this we have to add vegetables and fruit when available. We have records of onions, garlic, leeks, spinach and lettuce, plums, apples and cherries.

Two features stand out from this simple diet and what we know of its relationship to tooth integrity. First it seems to have been abrasive, for the chewing surfaces of all teeth in the skulls show progressive wear with age, and the fissures of back teeth disappeared in the late teens thus removing a major plaque trap. One cannot be sure how this came about, whether from millstone grit in the flour or the roughness of everything eaten. However it happened, it is not likely to be easily reproducible in any modern version. We shall in our recipes, however, be using 'whole' ingredients as far as possible, partly because, as far as the cereals are concerned, there are known preventive gains[14] but chiefly because they are tastier and contribute more pleasant textures. There are, moreover, medical considerations in the retention of roughage and fibre which we are happy to incorporate. Certainly we will avoid those genteel cookbook instructions to remove every pip, skin and husk from every rude vegetable, leaving every product like a blancmange smelling faintly of garlic.

The second feature of this diet was the absence of refined sugars, though a small amount of honey may have been included. This has to be the cornerstone of our diet and in constructing it we have an abundance of advantage that the medieval peasant could not have dreamed about.

Autumn meant the end of all the green food from woods and commons, the last wild fruit was finished, and the corn had been gathered (they cut it high by hand with reaping hooks, and the cattle browsed on the straw). The geese and hens had eaten the fallen grain and had been eaten in turn, the pigs had eaten the fallen acorns and beech mast, and now the pigs were killed off for winter, one by one. Winter came, and there were salt and dried meats, and some parsnips. After a good harvest these lasted well enough, but as the winter dragged out, the last of the fresh meat went; only the few beasts necessary for breeding in the spring remained . . .[15]

Against this threat, technology has long since set up formidable defences. Seasons have to a large extent been abolished by international trade and the freezing and canning industries. More sophisticated nutritional knowledge, while a mixed blessing, has enabled us to extinguish any serious possibility of nutritional deficiency. Tastes have expanded to admit greater variety of food than the common round would hitherto allow. We are richer and do not have to spend our whole days in keeping body and soul together. With our riches we can buy leisure, labour-saving devices and newer ways of preserving food such as deep-freezing and irradiation. The notion that food and its preparation can be among the great joys of living is coming into circulation again and new cookery books appear daily to enrich our culinary horizons. It remains to be seen, however, if the wise options contained in this armoury of advantages will be taken up, for the books themselves are still rooted in the sugar culture.

We have to start our dental health diet with a decision not merely to cut down on the use of bag sugar but to remove it entirely from the scene. This is a necessary baseline forced upon us by the insistence of the commercial pushers that it should intrude into every recipe and ingredient list like a hiccup. We also set our limit of sweetness at that of the fresh fruit that would have been our medieval ancestor's ultimate, and achieve it by a blend of naturally sweet ingredients.

We will not of course smuggle sugar concentrates in through a honeyed and syruped back door nor will any of the recipes use any artificial sweetener. High levels of sweetness from whatever source stifle delicacy and subtlety of flavour.

We do, however, list commercial food and drink in which artificial sweeteners are used. This is for the convenience both of those who wish to avoid them and those who find a total ban on occasional sweet indulgence unacceptable. The only ingredients used that come near the status of concentrates are dried fruits. These, however, are sparingly used to raise the palatability of blander ingredients, but only to the limit set above, i.e. the fresh fruit level.

In this way the overall consumption of sugar from all sources is bound to fall and the gap will be filled by dietary elements of broader nutritional value. These can include foodstuffs which exert positive dental benefit. We are not here talking about those hoary old-timers, the apple and carrot, which were supposed to clean us up after the treacle pud. Unfortunately this tradition has been rumbled. We now know that not only do apples *not* remove plaque but that,[16] taken after a dessert course, they will actually increase its acidity – the end of the apple as nature's toothbrush, if not quite of civilization as we know it today! It need not be the end of the apple, of course, only of its place in the meal. Indeed, together with the carrot, it remains one of our best snack foods.

The apple's drop in the charts does, however, lead us back to a consideration of 'acidity' in relation to tooth decay, not just that caused by the sugar but the intrinsic acidity of the foods themselves. It is an issue made much of by the sugar salesmen and enables them to talk loftily of dental decay being a multifactorial disease, i.e. caused by many factors. The word is brandished about so proudly that it almost seems unkind to point out that the main cause by far is still sugar amount and frequency and the trickle of factors justifying the multi prefix is of relatively minor significance. Many of our natural and manufactured foods have a natural acidity well above that at which tooth enamel can dissolve. This is not really the point. What we need to know is the acidity of the *plaque* after the food has been taken and how long it lasts. We know for instance that one effect of strongly flavoured or acid food is to stimulate a fast flow of saliva.[17] The faster this saliva flows the more alkaline it is and the more it swallows up and buffers the acid. After all the acidic vegetables and fruits eaten by our Anglo-Saxon ancestors took decades to make their gnawing presence felt. They kept their teeth most of their lives in spite of having no access to toothbrushes, magic pastes and fluoridated water.

They had further advantages in addition to the absence of sugar. Their cereals were unrefined. We now know that in the outer layers of oats, wheat, rice grains, pecans and peanuts there are protective factors, though it is doubtful if their effect could offset the massive sugar intakes of modern man. These will however play an important role in our low-sugar diet in combination with the natural sugars of fruit and vegetables. Protective factors are also present in cocoa and liquorice. The former will also be involved in our recipes for sweeter food while liquorice allows a pleasant exception to the sweet limitation.

In addition to the modifiers which limit plaque acidity, we now know of certain foods which cause it to drop rapidly. Of those studied the most important are cheese and salted peanuts.[20,21] This effect clearly has important consequences at mealtimes. The traditional advice to make for the wash basin at the meal's end is not one that is commonly followed, or if it is, the acidity has usually been sustained for some time before the brushing starts. Less anxiety (and perhaps better digestion) would be caused if this dash to the basin was replaced by taking a small piece of mature cheese after every dessert course. We could even improve on tradition by leaving out the dreary biscuits which so often accompany it. The French in their conservatism might still prefer to proceed from entrée to cheese in order to finish off the claret more pleasantly. In this case they would be wise to leave it at that and forget about dessert course and sugary coffee or liqueurs.

No culture known to us supplies salted peanuts as a final meal course and it is difficult to see how this powerfully acid-reducing ingredient can integrate with a meal. Furthermore, excessive salt intake is thought by nutritionists to be a general health hazard. Peanuts will be used in the next chapters, however, both as snack and butter in our battery of alternatives to the between-meal sugar-weasels.

We are now therefore on the way to assembling the battery of food choices against the invaders. There is much on our side. There are hints in current research that other factors militate in our favour. On waking, our plaque acidity is often lower than that of the saliva which bathes it,[22] which indicates that the plaque by itself is not the arch-demon that the reform church insists on; that left alone it can exert positive benefit. It contains high concentrations of substances like calcium, phosphate, fluoride, and proteins, which are all protective against acid attack. Moreover, between the plaque and the enamel surface there is also a thin film (called the pellicle) which may also be protective. Saliva, too, contains a substance, not yet fully studied, which exerts an acid-reducing effect.[23] It is the juggernaut thrust of the sugar culture with its ubiquitous incursions into our lives which frustrates biological control. Since dietary habits are formed less from rational adult choice than childhood experience we begin our strategy at the beginning, in pregnancy.

4 The milky way
Infant feeding interferers

The first thing necessary for success
in life is to be a good animal.

Herbert Spencer

She gave her sonne suck and tooke
great comfort and delight therein.

The Countess of Lincoln's Nursery

'My little boy/girl is two/three/four. When shall I start bringing
her/him to see you?' The question is probably asked every day in
every dental surgery and it is a wrong one, or at best a very
secondary one. The primary question should be, 'I am expecting a
baby. What can I do to prevent it getting bad teeth? How can I
keep my baby *away* from the dentist?' For if this mother waits
until the child is two or three before thinking about teeth, all the
dental visits in the world will not alter the dietary habits which by
then will have become deeply established.

This mother will have been, in the months of pregnancy and
nursing, subject to powerful forces of commerce, folklore, and
occasionally, alas, the nutritional ignorance of medical personnel.
She may well have been sent to the dentist for a check on her own
teeth during this period. It would be exceptional if this resulted in
a dietary scheme for the child, however skilfully repaired her own
teeth might become. It is also rare for ante-natal clinics to offer
much about dental disease prevention, maternity homes even less.
In the latter, commerce, with its free samples of syrups and baby
foods, penetrates within days of the birth. These are accompanied
by free booklets on baby care, sponsored usually by food manufacturers whose concern is far from disinterested. By this time
folklore, buttressed by commerce, has deflected dental interests

towards anxiety about lack of calcium and vitamins, all of which are redeemable by mere purchase. The trio of injunctions discussed in our first chapter will by now have become transmuted by a familiar alchemy into the following duo. Firstly, when the baby gets some teeth he or she must be got used to the dentist, and secondly, it doesn't matter much about eating sugar provided her or his teeth are cleaned regularly.

So when this child reaches the dental chair he or she may already have twenty teeth, some of which are decayed. Her or his eating habits are possibly immutable and the mother is already convinced that, having made the effort to deliver the child to the chair, her responsibility is exhausted. The dentist, if preventively oriented, must view the scene with a sinking heart, as yet another innocent mounts the treadmill of repair dentistry.

To revise this depressing agenda we start in pregnancy. The body's own magic is likely to be a fairly good guide to what a mother should eat. If she has a sweet tooth she should satisfy it with caution and read this book to the end. This, plus a bit of sensible weight-watching, should suffice. The simple tooth-cleaning measures described in Chapter 13 should be followed carefully but unobsessively as there is some evidence that an improvement in a pregnant mother's oral hygiene can give her child some immunity *before it is born*![1] This beguiling possibility apart, it is the dietary habits that begin at birth that will determine the baby's dental health, and these must be planned before birth. There is no conflict here, incidentally, with general health considerations. A diet good for teeth is a diet good for general health also.

At first it is simple. There is a single product that alone will keep a baby healthy and happy for at least the first four months of its life. This is the milk produced in its mother's breast. It is a measure of the corruption of sensible eating habits by commerce that this still has to be argued. This milk is custom-built, has the right temperature, its composition is automatically programmed, and the proteins it contains will not give rise to allergies. Breast-fed children tend to have less stomach and bowel trouble, and given free access to the breast, control their intake according to need. The breast-feeding mother will therefore be less anxious, and as a bonus will get her figure back more easily, without the misery of a slimming diet. Both parties are emotionally enriched, bonded, and pleasured by the experience.

The reported decline in the practice of breast-feeding has then nothing to commend it from a nutritional point of view. Studies have shown that at least 95% of mothers *can* feed their children.[2,3] They show also that, depending mainly on where they live, only between 15% and 30% are doing so at three months.[4] There are signs that the trend is now in reverse but it is still worth examining reasons for the original decline.

Much of our culture stems from a Judaeo-Christian tradition which allots to man a status somewhere between God and the other animals. This leads quite naturally to, among other things, an easy estrangement from our animality, to a horror that so spiritual a creature as man should be so earthbound by the necessities of animal function. Thus it can seem that we are demeaned by sex, sweat, defecation, menstruation, indeed by the whole of our physical nature. The 'bottle culture' has offered deliverance from this. White-coated science has assured the nervous mother that she can be protected from shameful exposure without threat to her child's well-being.

This prudery is not without its aggressive component. Since the bottle-feeder became the norm, the breast-feeder has become the crank who probably even makes her own bread! The practice of breast-feeding can in some circles be safely and sneeringly referred to as a 'cult'. A mother who looks for minimum tolerance and occasional facility for feeding her child away from home risks being regarded as over-demanding, a nuisance or even an exhibitionist, while her fellow who wields the bottle on the park bench will be rewarded with indulgent smiles from every passer-by.

Almost everybody is concerned that the bottle-feeder should never be allowed to feel guilty, a state of mind universally acknowledged among the enlightened to be 'a bad thing'. But little serious thought appears to be given to the multitude of anxiety-creating forces which can frustrate the breast-feeder. A mother's milk production is intimately circuited with her nervous system and this can be upset by many factors. In addition to the prudery discussed above, the mother, before and after giving birth, is faced with pressures from both commerce and professionals which often seem to be in rich collusion.

It would be surprising if baby food manufacturers were completely detached in their approach to infant feeding. However, a casual reading of their advice booklets might easily suggest a lofty disinterest. In them mothers can be informed that breast-feeding is natural, that the baby is less likely to develop colic, that it is satisfying. They might almost be, contrary to self-interest, in favour of mother's milk. That they are not intent on commercial suicide, however, is a fact shown by closer inspection when a garnish of qualifications emerges. For instance a check-list for *breast-feeders* on entry to the maternity home can include 8oz bottles with teats. The mother will learn that manual expression of milk should be taught so that she need not be tied to the baby for every meal; that breast-feeding does not come naturally, and nursing staff may be too busy to help; that instruction on bottle sterilization should be given to all since 'mothers may want to give a bottle feed occasionally'; that breast milk *provided that it is of good quality* can supply all the baby's needs.

Such subtlety of subversion is by no means the prerogative of commerce. Well-meaning professionals with no obvious axe to grind are able to pick up the anxiety baton without effort. Ask any midwife, health visitor, doctor or ward sister if they believe in and encourage breast-feeding. The answer is always an (at first) unqualified yes, coupled with a 'however could you doubt it?' amazement. In practice, however, the qualifications come in thick and fast and all of them spring from entirely worthy motives. They usually start with concern about the mother's fatigue after the effort of giving birth, that she should not be too taxed until rested. Meanwhile there is a baby to care for who may in the first 24 hours, especially if premature, be in some danger of having a low blood sugar. To get over this, a supply of 5% glucose is always available in ready-to-feed teated bottles which can be fed to a baby cradled in the arms of a devoted nurse. The mother may not immediately set up an adequate milk supply of her own, so that the baby remains unsatisfied. The question then arises (ever so quickly) as to whether 'she has enough'. To protect her from anxiety as the baby cries and to give her time to produce this 'enough', a supplementary feed is prescribed. This is a modified cow-milk of impeccable purity to keep the baby going.

Alas, too often the 'enough' does not arrive. The mother is then surrounded by deep and genuine sympathy for her disability. She is comforted and finally assured that she need not feel guilty. It becomes part of her personal folklore that she 'didn't have enough milk'. The hospital's reputation for the encouragement of breast-feeding remains intact, and everybody gets back to the 'normal' rituals of bottle-warmings, ready-to-feeds, and sterilizations, which, as it happens, are convenient for ward routine. Nurses who love little babies then have ample opportunities to cuddle and feed them while the mother gets a good sleep.

We therefore have to insert into this scenario a little lesson in human biology. This lesson cannot fail to have been taught at some stage to all the health professionals listed above. It seems however to be a lesson which easily succumbs to maternity ward folklore. This is that the breast is a gland which performs in response to demand. Don't use it and it stops working. To quote one famous paediatric source, 'the best galactogogue or stimulant of milk production is suckling'.[5] Breasts do not fill up or not according to a four-hourly bottle routine and the question of having enough is not a real one. The sooner the baby is put to the breast after birth the sooner the milk supply will be established.

It follows that the less the professionals interfere with this natural animal process the better. Left to themselves and given unrestricted access to each other most mothers and babies will sort out questions of need, supply, and demand. Concern for hospital routine should have no kind of priority in this matter. A hawkish paediatric eye must always be focused on the scene to pick up the emergence of unusual conditions but this ought not to be used as a camouflage for the shallow thinking about infant feeding which often dominates among professionals.

We are not, however, intent on setting up as breastfeeding counsellors. Our purpose here is to identify the hidden enemies which frustrate self-help. Any mother who wishes to breast-feed and finds poor support from her environment or the professionals should write for help to The National Childbirth Trust, 9 Queensborough Terrace, London W2, or La Leche League of Great Britain, BM3424, London WC1 6XX enclosing an SAE.

All this having been said, most babies are bottle-fed and they mostly thrive. What goes into the bottle is usually modified cow milk. The old-fashioned trick was to heat this to kill the germs, dilute it with an equal amount of sterile water and add sugar. There are obvious dangers in using a product designed for a baby which is trotting around within minutes of birth for one who spends most of its time asleep, and there are also risks in any procedure where measurement is important. The temptations to 'enrich' a feed with extra sugar, or to get the baby on to solids early by slipping a little cereal into the bottle, are constant and full of hazard. Such practices can lead to obesity in adult life, and a demand for excessive sweetness levels after weaning. Fortunately manufacturers have 'humanized' cow's milks to an extent that in all important respects they resemble human milk. Since sucrose is not used in their manufacture the sweetness level is also similar. Those currently available are: 'SMA', 'SMA Gold Cap', 'Cow and Gate Babymilk Plus,' 'Cow and Gate Premium', and 'Osterfeed'.[6] Instructions for these are clear and, water apart, nothing must be added to them.

Keeping to breast or humanized milks alone for three to four months will ensure provision of all the nutrients the baby needs.[7] The pressures of commerce and folklore, however, will not abate. Consider the following scenario.

Suppose you wanted to travel from London to Edinburgh and asked a travel agent for advice, and his reply was as follows, 'You ought to take a train. You find these at railway stations. Choose one with wheels. It should always have a driver, seats, roof, windows and doors. It should carry enough fuel to last the journey, so that it doesn't stop between stations.' You might find this advice a little odd, and wonder perhaps to which terrifying alternatives it supplied correction. But suppose, having set out on the journey you discover that your train is staffed exclusively by officials with raging infectious diseases. Such an imbalance of concerns you might regard as criminal or at least excessively batty.

As a metaphor for what comprises much dietary advice to nursing mothers it is by no means unfair. A mother, whose milk cannot fail to contain all the dietary elements necessary for health, is badgered constantly not to be without the marvels of modern science. Chief of such marvels are the vitamins, which like the train with wheels are difficult to avoid. With them, however, can come infection. They are the smokescreen behind which absurd levels of sugar are maintained in the diet. Since such a high noise-level of concern is raised about vitamins, some attempt should be made to put it into perspective.

Until this century it was thought that diets would be adequate if they contained enough protein, fat, and carbohydrate, plus a few minerals. It was then found that natural, unrefined foods contained substances essential to life and health which the body could not itself piece together out of such a diet, or at least not in sufficient amount. These were given the name 'vitamin', and their discovery completed to a large extent our knowledge of the relationship between food and health.

The name was an unfortunate one. The original expression, 'accessory food factor', however clumsy, might, had it survived, have saved us a great deal of trouble. As it happened, the earliest discovered factors were 'amines' and, since they were found to be vital to health, the name 'vitamine' was coined. All that had to happen then was the dropping of an 'e', and science's greatest gift to Madison Avenue took the stage. No longer was it a mere accessory food factor; the very word immediately suggests our old friends 'zip' and 'zest.' We have only to look at it to know that if we swallow enough we can take off into space like Superman. All that then remained for the ad-man was to scan the nutritional scene for the slightest anxiety about vitamin availability in normal diets, and then make his poetic most of it.

The colder truth is that once there is enough vitamin in the diet to prevent deficiency disease, adding more has no effect. In the case of vitamin D a large excess may even be dangerous. It is important to remember that vitamin C deficiency, at least, is now rare in affluent Western countries. If a child has it, the mother won't get near her or him for the photographers.

In a famous experiment on himself, in 1940, an American scientist carefully deprived himself of vitamin C in order to test whether scurvy and vitamin C deficiency were the same thing.[8] It took 40 days before his blood was cleared of it, and four months to develop signs of disease, and he really was trying hard. Vitamins are actually rather hard to avoid. Only the very ill and alcoholics who take no solid food tend to be deficient. The only other group to suffer on any scale is children who for reasons of climate combined with poverty suffer vitamin D deficiency and the possibility of rickets. No one, as yet, has come up with any serious link between vitamin deficiency and tooth disease. Nevertheless rosehip and blackcurrant syrups are constantly peddled as necessary aids to dental health against an encyclopaedic background of certainty that sugar is the chief criminal in dental disease.[9]

Commerce is yet again well-buttressed by the naivety and plain ignorance of medical personnel. Young mothers are bullied and blackmailed at post-natal clinics to buy expensive syrups, on pain of their children developing serious vitamin deficiencies. The dental health objection is met in two ways; firstly that the glucose syrup now used in the formulations is less cariogenic (decay-causing), and secondly by the injunction to dilute with an equal amount of water. Our reply to this is that although glucose is marginally *less* cariogenic than sucrose it is still *very* cariogenic. The dilution advice also takes us nowhere since it still leaves healthy teeth being bathed in a 30% sugar solution, perfect plaque food.

Even if we assume that evolution has somehow gone astray, that breast milk can no longer be regarded as an adequate source of nutrients, there is still no case at all for the syrups. The same clinics which peddle them also have sugarless drops containing sufficient quantities of vitamins A, C, and D. They are much cheaper (free to some mothers and currently 18p for a 10ml bottle in the UK) than the syrups and will not supply a relatively inactive baby with surplus calories. Other than these, and, in cases of thirst, a spoonful of sterile water, nothing should intrude upon the hegemony of milk.

Given that the sugar-pushing is not easily resisted by medical personnel, the results of a recent survey of health visitors and mothers are worth an ironic mention. They were asked for reasons to give a child sweetened food and drinks. Only 12½% of the visitors said there weren't any, as opposed to 40% of the mothers.

One further anxiety needs to be discussed in warding off these potential intruders, and that is the 'calcium-lack syndrome'. There *are* conditions such as vitamin D deficiency where our inability to *use* calcium is impaired but it is uncommon for calcium itself to be deficient. No scientific link has been established, however, as with vitamins, between any lack of the stuff and weak teeth. In any case there is a lot of it about, and we usually take in more than we need and have a sensitive mechanism to control the amount we keep. Even the breast-feeding mother who can lose a third of a gram of calcium a day through her milk is rarely at risk, though if anyone should suffer it is she. If she is not taking enough in her diet, her bones (never her teeth) will be depleted to keep the baby's supply going. In general the baby gets the choice courses and the mother the dog's breakfast.

Anxiety about vitamins and calcium is good for manufacturers but rarely for the rest of us. Its chief danger is that it deflects our attention from the sugar which creeps up on our lives from a wide variety of sources. One source, of profound psychological importance, is unfortunately enough the medicines themselves. The same 66.6% sucrose syrup which makes up the rosehip tooth-rotting preparations is used as a vehicle for children's medicines. This use did not start with Mary Poppins, though the lady surely has a lot to answer for. Most medicaments a few hundred years ago tasted pretty foul. The intense sweetness of honey, then syrups was employed to submerge this repellence. The tradition has carried through to a further usage. Children's dosage of the more powerful modern drugs is much less than that required for adults. Rather than create an expensive range of smaller pills and capsules, the manufacturers prefer to disperse the drugs in liquid to enable precision of dose. To prevent spoilage and contamination this liquid is nearly always a sugary syrup. Thus any doctor wise enough to be concerned about the dental dangers of sugary medicine would find it difficult to prescribe for children without recourse to them. It is not merely that dental disease is thereby actually *caused* by medical personnel, but that the enormous weight of medical authority is seen to support the *necessity* of sugar in our lives.

Given the will, and possibly public and professional pressure, the drug companies *can* solve the problem. In February 1983 dentists received a circular from Bencard, the makers of the anti-biotic Amoxil, advertising a 'sugar-free Amoxil preparation' especially designed to protect children with valvular heart disease who have to have dental treatment. In September 1983 Allen and Hanbury's reformulated their Ventolin syrup to an orange-flavoured but sugar free preparation. The principle should surely be extendable to other products. Meanwhile a major problem remains.

The study already referred to in this chapter has shown that 12% of children under a year old are taking prescribed syrups for periods of two or three months at a time.[10] This authority is shored up by the habit of using sugar lumps as vaccine vehicles and sweets as 'reward' for bravery. The sweets most commonly distributed closely resemble the pills which posters in the same room warn a parent against leaving around. Health visitors have been known to use sweets as 'grasping objects'. There is in fact a long tradition of medical indifference about teeth, and it is not so long ago that the mouths of teenagers could be cleared at the first hint of gum-bleeding, or because of a faddish notion that they harboured a 'focus' of infection. The mentally ill in the 1930s were peculiarly vulnerable to this practice. With no scientific basis whatever, wholesale extractions were often performed on the psychotically ill on the presumption that teeth were the cause of the illness. One pyschiatrist reports that in the North of England and the Midlands in the mid 1930s there were whole hospital wards without a single tooth. And if this seems an unfair attack upon a speciality which has now put its house in order, we willingly concede which profession it was that did the actual pulling.[11]

Dealing with medics on an individual basis is unlikely to have any success at all. Even the most approachable and unstuffy doctor can do little if the tradition is so insistent. Individual pressure can and should be applied at a later stage for non-sugary alternatives to be prescribed. Until this can be done each dose should be well-diluted and, for reasons explained in Chapter 3, followed by the chewing of a small piece of cheese.

Caution should also be exercised in the chemist's shop over unprescribed medicines. Gripe water, in particular, is an unpleasant sugary mixture for whose effectiveness there is little evidence. The persistent colic pains which some children experience are a matter for medical investigation, not an old fashioned remedy.

The only other expedient that might be required in the first six months is a teething preparation. Some of these contain large amounts of honey. The sugar-free ones are: Anbesol Flavoured Gel, Bonjela, Hewlett's Teething Jelly, Maw's Soothadent, Moore's Teething Jelly, Steadman's Teething Jelly and Oral-B Dental Gel. If teething pains are affecting appetite, the gels should be applied before meals. They contain numbing agents so it is best not to be so liberal with them that the baby's air and throat passages are anaesthetized, thus affecting cough and swallowing reflexes. In extreme cases a general painkiller, such as soluble baby aspirin, should be given by mouth, dissolved in a spoon of boiled water or very dilute fruit juice. The child should then be put straight to breast or bottle.

Finally, just within the 'milk phase' a baby may be grasping objects to suck and gnaw. Teething objects and rusks are a pathway along which the sugar culture can enter a baby's life and discussion of their role is important. This discussion will be more appropriate to the next chapter where a start is made on all solid foods.

5 Towards solid good
Weaning wisely

Come, bombs, and blow to smithereens
Those air-conditioned, bright canteens,
Tinned fruit, tinned meat, tinned milk, tinned beans
Tinned minds, tinned breath.

John Betjeman

A baby at four months should have been fed exclusively on milk,
whether from breast or bottle. The only safe addition to this will
have been vitamin drops. The sensible parent will have resisted
pressures from commerce, nosy neighbours and know-all grannies
to supplement the baby's diet with anything else. In some cases a
mother will find that breast milk alone will keep her baby content
to six months. In others, when perhaps the baby is large (not fat
but long and broad) and demands very frequent feeds, there is a
case for solid additions at four months. Early solids cannot be
regarded as a panacea for the various ills of colic pain or sleep-
lessness, however.

It is important to emphasize that there is no other rule about
when to stop breastfeeding than that decided between mother and
baby. In this respect professional advisers tend to be on the
defensive, hoping for at most two or three months of exclusive
breastfeeding before capitulating to the junk-food colossus which
insistently intrudes upon what ought to be a private affair. The
only 'normality' is what mother and child decide for themselves. A
mother can feed her child for years and, in less squeamish cultures
than ours, often does with great nutritional benefit. It is no business of health professionals, neighbours or relatives to interfere or
impose a time limit, whether this is done by direct advice, a shake
of the head or a raising of eyebrows. Only when mother or child
begin to feel they want a change should the question of weaning
arise.

Milk is a very balanced food and what the baby is weaned on should eventually have similar balance. The word *wean* is derived from the Old English *weinian*, to accustom. Although it has other senses, this is the one we want to stress, for the management of this phase may be crucial to the baby's dental health. It is what the baby is 'accustomed to' in these months that will largely determine its preferences throughout life. If variety is established in these early months the possibilities of faddishness and opinionation are much reduced and the options for a dentally healthy diet much increased. Since milk remains the main provider in early weaning it follows that balance is of minor importance as compared with variety. The burden of this chapter is to show how this variety can be achieved.

First a word about drinks. Teeth normally start to arrive in the sixth or seventh month and are immediately vulnerable to bad practices. In an earlier chapter we explained how a sugared 'drip-feeding' of plaque could keep acid levels on tooth surfaces dangerously high. In ordinary practice such feeding is found when a bottle is 'propped' against the child as a comforter, or when 'dinky' feeders are used. Since such mother-substitutes are commonly filled with syrupy vitamin preparations or sugared milk, trouble should surely be expected and it certainly comes. There is a special form of 'rampant' tooth rot associated with these habits.[1,2] Even breast milk is not entirely safe in certain conditions. An American study[3] of the children of four hardline breastfeeders showed that the constant day-and-night drip feeding had brought about widespread tooth damage. Circumstances here were very unusual: the mothers were in a state of extreme anxiety about husbands serving in Vietnam, and much more was asked of the practice of breastfeeding than it ought to be expected to supply. Ordinary demand feeding in the less fraught situations normally encountered offers no such hazards. There is, therefore, an important 'liquid weaning' which accompanies the move towards solid food. For the breastfeeder there need be no intermediate bottle phase. The transfer can be from breast to breast-and-spoon, to breast-and-cup and finally to cup alone. Bottle feeders would be wise to use the bottle only for milk, and wean to spoon-and-cup for other drinks.

The order in which sieved foods are given to a child is not important.[4] Traditionally the first food is a finely milled cereal product with additions of iron and vitamins. There are no sound nutritional reasons for this. True, the baby's liver store of iron is running out and the vitamin C level of breast milk is not high, but against this there is a danger of over-feeding a relatively inactive infant and making it fat. We suspect that the attachment of parents to cereal weaning has little to do with any sophisticated awareness of vitamin needs. Cereal is white like milk and this may make it seem appropriate. (Wholefood shops do a browner and very tasty version.) It would be far safer to raise the iron stock by starting with a teaspoon tipful of egg-yolk and then progressing to a whole yolk of lightly boiled egg. The white is best avoided at this early stage, because of possible allergy problems, but chopped up it will improve the family's mashed potato marvellously.

To make cool common sense out of weaning, away from the commercial din, we need to look at how nutritionists classify these solid foods. They have listed four main groups of food and claim that if something from each one of them is taken every day, all nutritional needs will be satisfied. This is helpful, though it is doubtful if many mothers are seen clutching such lists around the supermarket. We don't, after all, eat lists of food. We eat meals and their composition is largely a matter of taste and custom. We eat what we fancy and what we have been brought up to regard as normal. Nevertheless it is good to have a background checklist which can serve as a corrective if doing what comes naturally looks like making problems. The baby's diet, however, ought to be related to it and tastes should be allowed to develop over a wide variety of foodstuffs and not influenced by the parents' prejudices.

The groups are as follows:

The high protein group For meat-eaters this includes meat, offal, fish and eggs. For all, including vegetarians, nuts, soya beans, peas and other pulses.

The dairy group This is close to what the baby has been used to. It includes milk, cheeses, creams and yoghurts.

Bread and cereals Contrary to popular belief this is not just starch. Bread from wheat is rich in protein. Whole grain products are rich in the B group of vitamins while the husk containing them is valuable for its roughage. Some husks have an important dental role. This group is important because it is the cheapest.

Fruit and vegetables As well as carbohydrate this group provides mineral traces and a little protein, together with vitamins. The fibre bulk content is high. With certain reservations, it can be seen as providing the best and dentally safest of between-meal snacks.

To these groups must be added another which we shall call the concentrate or additive group. It contains the cooking fats and oils, as well as the sugar concentrates like sugar itself, honey, molasses and syrups. None of these is essential but they occupy a central place in most kitchens since their use makes food physically attractive.

At the weaning stage it is the fourth group that assumes major importance. Since fruits and vegetables figured prominently in primitive and dentally healthy diets we have to encourage expansion of their use and variety. There are also good economic reasons for this. We live in a time of world food shortage and in our children's time this is likely to grow more acute. As a result our dietary habits are coming under close scrutiny. The fact, for instance, that it takes seven pounds of vegetable protein to produce one of meat, is making people wonder whether it might not be a better policy to eat the vegetable in the first place, without passing it through an animal. Already the price of animal protein has caused caterers and food manufacturers to 'extend' meats by the addition of the soya bean-based TVP (textured vegetable protein). It is hard to imagine the trend being reversed.

We have, therefore, prepared a fairly comprehensive section on fruit and vegetable preparation, though of course, given the short time span of early weaning, it is not likely to be fully utilized. Here an electric blender is a most valuable tool, and can no longer be regarded as a luxury item, but a hand blender, though more labour intensive, will do the job just as well. With some fruits and vegetables even this will not break the material down to swallowable consistency and sieving must follow. There is no point in protecting a baby from fibre which it needs as much as any adult. Provided the food is swallowable and the baby cannot choke on it, it is acceptable.

One further point. If canned or frozen alternatives are used in this section an eagle eye will be required in the shops. Sugar has been weaselled into most canned fruits in the form of syrup and it even gets into tins of vegetables. Tinned vegetables often also contain salt, which should be rinsed away before cooking. Some of the frozen fruits have also either been dipped in syrup or sprinkled with sugar and must be avoided. Exceptions to this rule are several brands of canned pineapple in natural juice and others which are listed in Appendix 1.

'CONTENT WITH A VEGETABLE LOVE'

Asparagus

Plunge one shoot into 2.5cm (1 inch) of rapidly boiling water and simmer for 20 minutes. Drain and lay on a board. With a kitchen knife separate the pulp from the fibre by stroking it from stem to tip. When all that remains is the stringy residue, sieve the pulp.

Beetroot

Wash a small beetroot thoroughly, taking care not to prick or break the skin. Put into boiling water and simmer for 1½ hours. (A pressure cooker will cook it in about 20 minutes.) Cool in cold water enough to allow you to pinch up and peel off the skin. Mash with a fork or sieve.

Green beans (French, bobby or runner)

Wash, top and tail, and string if necessary. Slice diagonally into 3cm (1½ inch) lengths and plunge into enough boiling water to cover. After 10 to 15 minutes of simmering they should be tender enough to be pressed through a sieve or hand blender. The mush can be thinned with a little of the water. If you are using frozen beans there will be cooking instructions on the packet.

Dried beans

These really count as Group 1 foods but are included here because most people would count them as vegetables. Whether they are haricot, broad, black-eye, red kidney or whatever, they are treated the same. Soak them overnight and pour away the soaking water. Cover with cold water and bring to the boil. Cook rapidly for 10 minutes then discard the water. This practice is important in order to boil off certain poisons present in the darker beans which can do serious damage to the gut. Bring to the boil once again with fresh water and cook until tender. Press through a sieve or blender and thin out with a little of the cooking water. Remember tinned baked beans contain sugar. In Chapter 11 there is a recipe for haricot beans in tomato sauce which provides a very tasty and far superior substitute for baked beans. Sieved or blended this can well be served to a baby.

Broad beans

If fresh, shell. Add to enough boiling water so that they are just covered. Simmer 15 to 20 minutes. Sieve or blend and thin out with a little of the liquid. For frozen beans, cook as the packet instructs, then sieve.

Cabbage, broccoli, spring greens, cauliflower, sprouts

Wash thoroughly and shred. Add to 2.5cm (1 inch) of rapidly boiling water and immediately replace lid. Simmer 15 to 20 minutes. Mash with a fork, then sieve, adding a little water to thin.

Carrot

Scrub, top and tail, and slice. Add to 2.5cm (1 inch) of rapidly boiling water. Cover immediately and simmer 20 to 30 minutes. Drain and mash or sieve. Turnip and swede can be treated in identical fashion except that they need to be peeled first.

Celery

Remove leaves and wash. Chop and add to 2.5cm (1 inch) of rapidly boiling water. Cook with lid on for 30 minutes. Drain and sieve.

Cucumber

Yes, a baby will eat cucumber. You need only a small piece which you peel and halve lengthways in order to scrape out the seed. Grate first, keeping the juice, and then sieve, adding the juice later.

Lentils, dried peas, split peas

Wash and leave soaking in cold water for 2 hours. One tablespoon to four of water is about right. Bring to the boil and cook for about 45 minutes. Sieve and thin out with water.

Marrow, courgette and squash

Use either a slice of marrow or squash or a small courgette. Cut up and drop into boiling water. Simmer covered for about 15 minutes. Sieve. Don't worry about the pips; the sieve catches them.

Parsnip

Proceed as for carrot but the cooking time is shorter. About 15 minutes is enough.

Peas

Add the peas to 2.5cm (1 inch) of rapidly boiling water. Cover and simmer for 10 minutes. If using very fresh young peas or frozen peas, halve the time. Sieve or blend, thinning with the water if necessary.

Potato

Boil a washed potato in its jacket, or if you have the oven on for something else bake at 200°C/400°F (gas mark 6). Boiling takes just under 20 minutes, baking about twice the time. Run under cold water for a few seconds to make it easy to handle, peel and mash with a little skimmed milk.

Sweet pepper

If fresh, cut in half lengthways and remove white pith and all seeds. Chop and add to a little rapidly boiling water. Simmer for 5 to 7 minutes, then sieve or blend.

Tomato

To peel a ripe tomato, immerse it in boiling water for a minute or two. This actually cooks a thin outside layer and the peel comes off more easily. Sieve and serve.

FRUITY BEGINNINGS

When a fresh fruit is fed uncooked to a young child, it is at first more acceptable if slightly warmed. The pleasure of cold food does not dawn immediately on a baby used to milk at body temperature but if introduced gradually will soon be enjoyed.

Apple

Cut an eating apple or ripe cooker in half. Scratch out the pulp with a spoon and feed this direct. Treat a ripe pear in the same way.

Wash, peel and core either type of apple. Heat slowly in a saucepan with a tablespoon of water, i.e. just enough to stop the apple from catching. Depending on the sort of apple it will 'mush up' in about 5 or 10 minutes. Mash with a fork and sieve.

If you are stuck with only fairly sour apples, add a few raisins before cooking and strain when cooked. If you have overdone the water addition, strain it off and use for a drink, perhaps when the baby gets hiccups.

Apricots

The apricots most of us know are dried. If you are able to obtain and afford fresh apricots treat them as you would plums (below). Otherwise soak the dried ones overnight, then mash with a fork or sieve or liquidize. To be on the safe side you may wish to boil them up with a little water for a minute or two before mashing. You can then reserve the liquid for drinking.

Banana

Choose a ripe banana and simply mash with a fork.

Blackberries, etc.

Blackberries, bilberries, blackcurrants, redcurrants, whitecurrants and raspberries are all valuable sources of Vitamin C. The currants particularly are rather tart, especially when cooked, and are lacking in bulk. The flavour should therefore be introduced by cooking them with something else. The classic is blackberry and apple:

250g (8oz) ripe cooking apples   125g (4oz) ripe blackberries

Peel, core and slice the apples. Simmer with the blackberries and enough water just to cover the bottom of the pan, until the apples are tender. Liquidize or merely mash and sieve.

Blackcurrants

The tartness of blackcurrants is a factor that always leads the traditional cook to the sugar pot. This is quite unnecessary. A splendid purée made with a thinned-out infant cereal to supply the bulk still conveys a powerful blackcurrant flavour. The tartness of any fruit after cooking often provides a case for eating that fruit raw, never for sugaring it. Ripe blackcurrants strained through a mesh are enjoyed by a baby without any other treatment. But here is the recipe for the blackcurrant cereal. Blackcurrants are the richest source of Vitamin C among the fruits, a fact that manufacturers of sugary currant syrups never fail to push. What they are actually selling you is sugar at vast expense. Better to stick to where the goodness is, in the berry.

1 tbsp blackcurrants
4 tbsp skimmed milk

½ tbsp infant cereal (any non-sugared cereal like Farlene or those listed in Appendix 1)

Put all the ingredients in a pan, bring to the boil, stirring constantly. The stirring will gently bruise the fruit, permitting its flavour to permeate the cereal gradually. Turn down the heat once boiling starts and cook for about 5 minutes, still stirring. Sieve and serve.

Grapes

Choose ripe grapes, halve them, pick out the seeds and discard. Scoop out the pulp then sieve or mash.

Melon

Simply peel, remove seeds and mash or sieve.

Oranges, grapefruit, tangerines, satsumas

Peel, remove pith and pips. Chop and liquidize. You may need to add a little water to get a thinner purée. Alternatively use a lemon squeezer and introduce the flavours simply as a juice.

Plums and larger soft fruits

The treatment for plums of all kinds (and apricots and peaches) is the same. Wash, peel, halve and stone. Mash or sieve. Prunes should be soaked overnight, the stones removed, and the flesh sieved. There is no particular point in cooking them if they are of good quality but if they are still rather hard after soaking simmer them gently for 30 minutes until they are plump and soft.

Gooseberries, strawberries and rhubarb

Are best avoided in the first year as their oxalic acid content may interfere with calcium absorption.

CHEWING THE FOOD OF FANCY

Finally, at the age of six months, a child can usually grasp a rusk or crust, carry it to the mouth to explore it and, if the taste is pleasant, will suck it. Chicken bones or others with no sharp edges are safe alternatives. Bones from very young chickens or rabbits should not be given as their ends can become detached.

There are products available in shops and pharmacies called Teething Rusks. A better name might be Teethrotting Rusks since they usually contain sugar concentrates like honey.

Recently some manufacturers, conscious of public anxiety about sugar levels in baby foods, have taken steps to alleviate it. Farley's Low Sugar Rusk has a sucrose level reduced to 15%. However, when we include the other sugars, the total becomes 28%. Similar 'reductions' on the part of other firms show that these steps in the right direction are exceedingly timid. Bickiepegs are unique among the commercial rusks in being sugar-free.

Rusks don't of course help teeth to erupt. Their use is really to give the baby something to do, or to keep it quiet. They can be made quite simply by taking a slice of brown bread and cutting out a shape from its centre with a pastry cutter or tumbler and then toasting it lightly. Otherwise, if you have an oven on the go for something else, pop in a few rounds to bake; after cooling simply store in an airtight jar. There is no special value in the toasting. It simply hardens the breads so that the baby doesn't crush it to pieces as one end before it has had a good go at the other. If you would like to add variety to the rusks, try the following recipes.

Egg rusks

| 1 slice bread, 1cm (½ inch) thick | 1 egg |
| 2 tbsp skimmed milk |

Cut the bread in 2.5cm (1 inch) wide strips. Beat egg and milk together and pour into shallow dish. Soak slices of bread in this and place in a cooling oven to cook. Cool and store in an airtight jar. This recipe provides a good way to use up leftover egg wash.

Cheese rusks

| 25g (1oz) hard cheese (grated) | boiling water |
| 1 slice brown bread | |

Cut the bread into 2.5cm (1 inch) strips. In a shallow fireproof dish add enough boiling water to the grated cheese to make a runny paste, stirring all the time. Turn the strips of bread in the cheesy fluid and bake in a cooling oven. If you want them in a hurry, bake on foil on a baking sheet for 15 minutes each side at 150°C/300°F (gas mark 2). Cool and store in an airtight jar.

Other and simpler alternatives are stick-shaped pieces of cored and peeled fruit or vegetable. The harder foods are least messy and should be kept to an unswallowable size (i.e. bigger than fist size). Among the fruits, a quartered apple or pear, peeled and cored, will keep a baby happy for ages. Raw carrot, parsnip, celery, swede, sweet pepper, core of white cabbage or turnip, well cleaned, pipped and cored, are all ideal. If a stick of carrot or other hard vegetable is kept in iced water and given to the baby during teething it provides some relief. For whatever reason they are given, they all have this much in common: they are tasty and nutritious and, because they are not sweetened, will not lead your baby to expect sweetness every time its mouth opens.

6 Mixed blessings
Weaning's end

> Children should eat as much as like at a time. They will
> never take plain food more than is good for them; they may
> indeed be stuffed with cakes and sweet things till they be ill,
> but of meat plainly and well cooked and of bread they will
> never swallow one ounce more than is necessary – Ripe fruit,
> or cooked fruit – BUT NO SWEETNING, will never hurt
> them, but once they get a taste of suggary stuff and when
> Ices creams tarts raisins almonds etc., and all the endless
> pamperin is come – the doctor must soon follow.
>
> *William Cobbett*

In the second six months of the baby's life, milk plays a decreasing
role. The solid food replacing it must, therefore, effect a similar
nutritional balance and meals will gradually come to resemble
those of an adult. At first they will be purées, but should gradually
become more lumpy as the teeth take over the work of blender and
sieve. Amounts required will still be small and, though the baby
can be fed adequately on shuntings from an adult plate, some
independent preparation will from time to time be necessary.

Two simple expedients can help lighten the chore. Most maternity homes use ready-to-feed milks and dextrose solutions in small
jars which are thrown away. These jars are exactly right for sorting
meal-sized portions so beg a box of them. The jars should be boiled
up with detergent, cleaned, rinsed with boiling water and dried.

Then use the deep-freeze to store bulk preparations. The jars
can be removed an hour or two before the meal for defrosting and
warmed in a pan of boiling water. The lids should be removed
before warming so that they don't fly off! If no jars are available,
use empty (5oz) yoghurt pots, other plastic throwaways or ice-cube
trays. The contents should be emptied into a pan for warming.
Commerce, of course, provides a wide range of (pre-cooked) sieved

foods ready to serve. There are several reservations to be made regarding these. Here is one paediatrician's comment:

> They (pre-cooked sieved foods) are commonly fortified with cereals and milk powder and are highly nutritious. This can be a disadvantage, for it may lead to the infant getting more energy than he needs, gaining too fast and becoming obese.[1]

More serious from the dental point of view is the high level of sugar in the dessert foods.

Sugar is often the first ingredient in items like creamed rice pudding, chocolate flavour pudding and egg custard with rice and in other desserts it is usually among the first three. There could be no worse dental start than to rely on such products. These sugar levels are, from the manufacturer's point of view, important as a preservative. They are dangerous to children because they set a sweetness norm for an impressionable child which can become a habit. The first and most welcome exception to this practice occurred in 1983 when Heinz introduced their new series of added-sugar-free fruit purées, linking this with a dental health message.

Home preparation is certainly cheaper and not very labour-intensive. We include a few basic recipes for savoury purées which can be permuted to offer a wide variety of choice. Bearing in mind the importance of avoiding *over*-nutrition we set a pattern based on meat (or fish) and three (or more) vegetables. Most adult meals can be liquidized provided that salt and heavy fats are kept to a minimum. On the dessert side, we sweeten only with dried fruit to a sweetness level which matches that of a fresh fruit and no further. Pudding courses can be the simple fruit purées described in the last chapter, or liquidized versions of those in Chapter 10. The fruit stew recipes in this chapter can be adapted as various fruits come into season.

Mince mash
(makes about ten portions)

250g (8oz) minced meat (preferably lean)
1 small onion, chopped finely
250g (8oz) potato (approx)

125g (4oz) carrot, swede, turnip or parsnip
250g (8oz) green vegetable

Add sufficient boiling water to cover the mince and chopped onion in a saucepan and simmer for 1 hour. Into another saucepan put the cleaned chopped root vegetables. Add boiling water and simmer, covered, for about 5 minutes. Then add chopped greens and simmer for a further 10 minutes. Pour off the mince water and discard (this will take away most of the animal fat fraction). Drain the vegetables, reserving the water. Add vegetables to mince and blend using the vegetable water to achieve required consistency.

Fish mash

Almost any fish fillet will do, though in the early stages it might be wise to keep to white fish (coley, cod, haddock, John Dory, plaice, lemon sole). Tail ends are least likely to have bone remnants. Amounts are the same as for mince. Prepare the vegetables as above, but this time use two plates as a saucepan lid, with the fish and a little oil between them. After the vegetables are cooked, drain as before, probe the fish for bones, and add to the vegetables. The liquor from the steamed fish is likely to be adequate to give the correct creamy texture to the mash, but have the vegetable water in reserve. This can be given as a drink if unused.

Chicken mash

The problem with chicken and other poultry low in fat is dryness, as thousands of ruined Christmas and Thanksgiving turkeys demonstrate. Breast cooks in about ten minutes while legs take over an hour and, unless carefully wrapped in foil, it is unwise to bake whole birds. For smaller amounts, 'sweating' in a tightly sealed heavy pan is the method to use.

1 small chicken quarter (leg is more juicy)	125g (4oz) carrot, swede, turnip or parsnip
250g (8oz) potato	125g (4oz) green vegetable
1 small onion, chopped	1 tbsp vegetable oil
1 clove garlic	1 tsp chopped tarragon or parsley

Heat the oil in the heavy pan and slightly brown the chicken in it. Turn the heat down very low and add the garlic, onion, and herbs. Cover and seal with foil. The cooking time is about 45 minutes. After 30 minutes cook the vegetables as for mince mash, above. Turn out the cooked chicken and remove bones. Liquidize all ingredients using vegetable water and chicken 'sweat' to cream.

Liver mash

| 125g (4oz) liver (chicken liver is the most tender but rather strong in flavour. Lamb or calf livers are just as good.) | 250g (8oz) potato 125g (4oz) other root vegetable 125g (4oz) green vegetable 1 tbsp vegetable oil |

Simmer the vegetables as for mince mash, above. Slice the liver as finely as possible and fry quickly in a small amount of oil. Drain vegetables, reserving the liquid. Remove liver from frying pan and add to vegetables. Deglaze the frying pan with vegetable water and use this to thin the purée to desired consistency.

Given the baby's gradual loss of its own liver iron stock, animal liver provides a valuable source in early weaning.

Gooseberry and apple
(makes about 10 jars)

500g (1lb) cooking apples	350g (12oz) can evaporated
250g (8oz) gooseberries	milk
50g (2oz) sultanas	

Peel, core and slice apples. In a very little water gently stew the apple, gooseberries and sultanas. Liquidize. Cool slightly and add the evaporated milk. Blend thoroughly in a liquidizer.

Blackcurrants can be used instead of gooseberries.

Blackberry and apple
(makes about eight jars)

500g (1lb) ripe cooking apples   250g (8oz) blackberries

Peel, core and slice apples. Simmer with the blackberries and enough water to cover the bottom of the pan until the apples are tender. Liquidize.

CEREALS

Cereals are not good weaning starters but they become more important as the child becomes more active and able to burn up their calorie-rich nutrients. There is a small number of sugar-free products in the vast sea of very sweet alternatives. The safe ones have been carefully tailored to nutritional needs and are pleasant in flavour. A parent who finds this flavour insipid should avoid reaching for the sugar bag or honey pot; a child is quite happy with the straight mix. A list is included in Appendix 1.

7 The long day's journey
Child's play

. . . We never see the flower
But only the fruit in the flower; never the fruit,
But only the rot in the fruit. We look for the marriage bed
In the baby's cradle, we look for the grave in the bed not
    living:
But rising dead.

Norman Nicholson
Rising Five

How can I teach, how can I save,
This child whose features are my own,
Whose feet run down the ways where I have walked?

Michael Roberts
The Child

The last two chapters were designed to allow the child to experience gradually a happy world of tastes, which includes sweetness but is not swamped by it. At the same time other developments are at hand. He or she is beginning to walk, to make the first efforts towards speech and to explore the inexhaustibly new and fascinating external world, a world without obvious structure, timeless and endless.

The mother's world on the other hand is very much structured. Her time scale is a crammed schedule of demanding tasks. The temptation to buy time with sugary silencers will be enormous and the simple injunction to watch the sweets is going to seem a tiny voice speaking from a long way off.

The dietary management of a child's dental health therefore needs placing in a wider context. We are dealing not so much with rows of teeth in need of chemical protection, or the germ warfare of the mouth, but with a complete person whose mouth can be ravaged by a complexity of causes remote from the world of dental surgeries and electroded rats. It is, it seems, a comparatively new concept to regard a child as a 'whole' person. Traditionally the first five years of a child's life are thought of as probationary, a tedium to be survived by trapped parent and infant 'pre-person' until the proper business of education begins.

Baby teeth do not escape this tradition which ordains that nothing should be done in the first five years which might then falter when the child reaches the age of five. Both child development and dental health are hampered by this obsession with the future, thus the idea that healthy eating habits should be encouraged is resisted on the grounds that 'once they get to school, there's nothing you can do about it'.

But all years are years of development and these early years should be enjoyed for themselves. Aching baby teeth will be shed in the course of adult time but for the toddler the aching day is long and within it he/she has to live with them, eat and speak with them, taste and smell them.

Dental health cannot therefore be viewed apart from all other developments. Thus, when the harassed and overworked mother meets a child's demands for attention and stimulus with a lollipop, there are two related consequences. First the casual ruin of the mouth is perpetuated, and at the same time a dead gift is substituted for live interest.

In most parts of the country, however, this captive mother is not now alone. Throughout the past two decades the resources of the playgroup movement have been at her disposal. The separate imaginations of many isolated parents have been welded together into a huge collaborative culture of child activity. Those parents already involved in playgroup activities and viewers of programmes like 'Play School' will want no help from a dentist on how to occupy and stimulate a child cheaply and imaginatively. Many are now expert at making aprons out of plastic carrier bags and paint pots from washing-liquid containers!

Where we will intrude is at the general level in urging the acceptance of controlled mess in the happy home, as well as the involvement of the child in as many of the household chores as possible. Indeed one is tempted to think that it might be a better bargain if double sinks were available on the NHS, than all that

71

expensive mouth-scarring metal! Any two- or three-year-old can be deeply absorbed in splashing about with (plastic) crockery in one sink for up to half an hour, while the serious washing-up goes on in the other. By any standards this is marathon attention-span. Pseudo, and later actual, help with bed-making, dusting and vacuuming are other devices well known to the experienced parent as stimulating and fun. Closer still to health goals is involvement with the enjoyable process of food preparation. Whatever the activity, one of the background aims is to sustain lively interest and to intercept that boredom which often surfaces as a demand for food.

8 Foundation courses
Bread and breakfast

O breakfast! O breakfast! The meal of my heart!
Bring porridge, bring sausage, bring fish for a start,
Bring kidneys and mushrooms and partridges' legs,
But let the foundation be bacon and eggs.

A. P. Herbert

The institution of breakfast is one of the great British contributions to civilization. While the other meals can be a bit Third Division compared with those culinary front-runners France and China, few foreign visitors depart unimpressed by British breakfasts. That depressing experience known as the 'continental breakfast' certainly comes nowhere in the culinary stakes, whether based upon the constipating, quick-staling, crust-and-air concoctions the French call bread, or those curling slivers of sweaty cheese and cold sausage that lie in wait for the unwary on the Costa Brava!

It is true that the *real* thing in England may now be so much on the decline that it remains for some little more than a folk memory, but the idea is a good one and has much on its side nutritionally speaking.

Slimmers know well that most eating is best done earlier in the day when there is the likelihood of being able to work it off. From the dental point of view a child, or for that matter an adult, who is fed substantially at the day's beginning, is perhaps less likely to be militating for snacks in mid-morning. A decade or two ago, a combination of fresh fruit, porridge, bacon and egg, followed by toast and marmalade, seemed the right foundation to take the eater through the harshest winter day. If we look at the groups of foods shown in Chapter 5, it will clearly satisfy the idea of 'balance' which used to dominate nutritional thinking. Since the real problem is now seen as eating too much of the wrong things rather than merely getting enough of the right ones, this notion of balance has been dropped, leaving the food provider in some confusion. This confusion is not helped by those diet advisers who seem to interpret the sensible guidelines of the NACNE Report as meaning that *all* animal fat must go and that *all* dairy products unless utterly shriven of fat content should be shunned.

We suspect that, for the physically active at least, the occasional sausage, bacon and egg type of dish can still be a worthy treat especially when grill and poacher replace the frying pan. But, alas, here we find another hazard, perhaps more potent than that posed by the born-again dietician. Take our old friend the sausage which, within a length of animal gut, used to house a blend of the minced cheaper cuts, some herbs and seasoning, mixed with good bread. Try then this supermarket version:

> Pork, water, turkey, rusk, beef, starch, salt, soya protein concentrate, spices, sodium polyphosphate, dextrose, herbs, flavour enhancer (monosodium glutamate), anti-oxidants E301, E304, E307 (i.e. sodium-L-ascorbate, 6-0-palmitoyl-L-ascorbic acid, synthetic alpha-tocopherol), sugar, preservative E223 (sodium metabisulphite), colour E128 (red 2G).

Have this, say, with a rasher of bacon,

> Pork, water, salt, sodium polyphosphate, sugar, monosodium glutamate, E301 (sodium-L-ascorbate), E250 (sodium nitrite).

Then a lightly poached egg which through its pallid yolk reflects the provenance of its laying in one of those silent wooden Belsens which fringe our villages, neat bungalow at the gate and a disinfectant dip for the lorries. For our money it will take more than Dvorak on a cobblestone video to make that chemical cocktail seem like real food.

We should nevertheless be able to restore a nourishing centrepiece to our breakfast that is neither commercial junk nor so loaded with animal fat as to turn our sedentary bloodstreams into a flow of thick custard.

The answer lies first in quality. If we are to eat less sausage, bacon and egg, let what we eat be good. We can occasionally poach a free-range egg, having first checked that it has been ranging on something more nutritious than a gravel path. We can grill the trimmed bacon and, as with the sausages, run off the melted fat. There are no candles to make any more nor wagon wheels to grease, so the drippings are as much a dead loss now as when they increased the girth of a beast to earn an E.E.C. subsidy. Finding the quality is another matter, but in the present climate of interest in diet and health many journals and newspapers are fighting a campaign for real food. Those who provide the genuine article are more and more likely to get wide and free publicity.

So variety of and respect for materials should be remembered as much as time and money allow. For hardliners who want to leave the animal world behind but would like a hot breakfast, there is always the vegetarian fry-up of whatever vegetables are in season, mopped up by the breads described later in this chapter.

Another traditional feature of the cooked breakfast due for resurrection is the use of left-over vegetables, meat and fish. Bubble and squeak, kedgeree and fish cakes, can all be organized the night before and require minimal preparation.

The barrage of pressure from advertizers to regard all serious food preparation as inconvenient has borne its dreary fruit in the common reduction of this good meal into a bowl of 'cereal'. There must be several generations by now who think that the word is exclusively connected with something coming out of a rectilinear packet, plastered with bargain offers and lists of essential vitamins and minerals displayed by the concerned manufacturer. There is, however, a smaller list on all of them which has greater significance, the ingredients panel. A careful reading shows that most of the raw cereals have been processed with large amounts of sugar. Only four brands are free from this intrusion and they are made from nothing more than whole (grain) wheat. These are Nabisco's 'Shredded Wheat' in its various shapes, Quaker 'Puffed Wheat', Sainsbury's 'Mini-Wheats' and Sainsbury's 'Puffed Wheat', and they must be regarded as the only dentally safe ones. For the rest, no amount of injected cheap vitamins and iron can obscure the fact that the nutritional balance of the natural cereal has been distorted by that other cheap injection – sugar.

Two others might be allowed, except by hard-liners, for the sake of variety as they contain only small amounts of malt, one of the least dangerous of the sugars. They are 'Grape Nuts' and 'Energen Wheat Flakes'. Since all these cereals are usually eaten with substantial quantities of milk they can be seen as meals balanced in themselves, especially when complemented by fresh fruit or unsweetened juice.

The original British cereal, porridge, has contrived a kind of survival through the hot oat cereal. Hardly anyone, unless they have a solid fuel cooker, prepares porridge the traditional way from raw whole oat grains cooked slowly overnight. Porridge oats are most commonly bought in a rolled form, even from wholefood suppliers. These oats have been heat treated while being crushed flat between rollers and are therefore part-cooked. Preparation time is therefore at convenience food level, a matter of a few minutes.

Such oats are not 'refined' and only the hard fibrous husk is removed. There may well be, then, good dental advantage here since, as we saw in Chapter 3, there are modifying factors in whole grain cereals that can mitigate the effect of sugar on plaque. There is no reason of course to confine our cereal interests to *porridge* oats. Medium ground oatmeal takes only slightly longer to cook and requires identical preparation. The larger oat flakes make a porridge of more chewy texture and the cooking time is about ten minutes. Wheat, barley, and rye flakes behave similarly. Wholefood shops supply a wide range of such cereals and often have recipe leaflets which explain how to prepare them. A preliminary soaking in the milk/water, perhaps from the time the kettle is put on for the first tea, brings the preparation time into the convenience range.

The instant porridges of the hot oat cereal kind are chiefly oat flour and may offer a little saving of time, but this cannot be very significant. However in spite of being pre-sweetened with a little malt extract they should be admitted to the menu for the sake of variety.

They include 'Ready-Brek' (plain), Quaker's 'Warm Start', Sainsbury's 'Instant Hot Oat Cereal' and Waitrose 'Instant Porridge', but not Kellogg's 'Extra' which is sugared.

CEREAL PLUS

A rather more expensive alternative is now invading the British breakfast scene, the mixture of cracked (unprocessed) cereals, nuts, and dried fruits known as muesli. Unfortunately most of the commercial brands, including those sold in so-called 'Health Food' shops, contain either soft brown sugar or some honeyed element. Exceptions are marketed by Cheshire Whole Foods, Just Naturally Foods (No Added Sugar version), Life and Health Foods (Norwich), Familia (Swiss Birchermuesli only), Waitrose (Fruit and Nut), Sainsbury, Boots (Second Nature), Tesco (Wholewheat Muesli) and Kellogg's (Summer Orchard). Most wholefood shops sell their own mixtures as well as a muesli base containing the cereal fraction to which dried fruits and nuts can be added according to taste.

Our own view is that the best mixtures are those with a large variety of ingredients and it is perfectly easy and certainly cheaper to put the mixture together at home. Granola, our second recipe, is a cooked version of muesli. Again commercial brands tend to be honeyed or otherwise sugared.

Muesli

2 cups rolled oats
1 cup wheat flakes
1 cup barley flakes
1 cup rye flakes
½ cup dried fruit (currants, sultanas or raisins)
2 cups bran

½ cup millet flakes
½ cup wheat germ
½ cup desiccated coconut
2 tbsp sunflower seeds
2 tbsp sesame seeds
1 cup mixed nuts (peanuts, cashews, hazelnuts etc.)

Shake up all ingredients in a big box or jar and serve with milk. If you like, sprinkle with granola or chopped fresh fruit. This is a basic recipe which can be varied according to taste or ability to obtain the ingredients.

Granola

1 cup barley flakes	desiccated coconut, wheat
1 cup wheat flakes	germ, bran, sesame seeds,
¼ cup each sunflower seeds,	chopped dates and oil.
soya flour,	

Mix all dry ingredients, blending well. Add the oil and mix thoroughly. Spread on a baking sheet and bake at 170°C/325°F (gas mark 3) for about 40 minutes or until golden brown, stirring every ten minutes to ensure even baking. Allow to cool and store in an airtight container.

SAFE AND SOUR

Another continental borrowing is the group of soured dairy products. These originate in countries where it is difficult to keep milk fresh because of the climate. Sometimes the temptation to make a virtue out of necessity has led the proponents of yoghurt, keffir and smetana to make extravagant claims for them as life-prolongers. Curiously, the consumption of yoghurt in particular seems also to give its consumers a sense of moral rectitude. If these strange attitudes lead to their substitution for sugared alternatives then this is all to the good. Nutritionally speaking they are, however, no better than the milk from which they derive. The home-made variety using skimmed milk also crosses no health picket lines. It is of course the product labelled 'natural' that is dentally safe and is available in most supermarkets. The yoghurt from goats' milk is likely to be found only in wholefood shops. The 'fruit' yoghurts are usually sugared to a high degree, as are some of those labelled 'low fat'. It is a simple matter to chop fruit and mix it in at home to make for a little variety.

USING YOUR LOAF

I NEVER NEVER liked brown bread
Whatever aunts and uncles said.
In vain they tried to make me see
The beastly stuff was good for me.
Though full of nourishment (said Nurse)
It looked like mud, and tasted worse;

A. P. Herbert

. . . American GIs have better food, live in more comfortable
quarters, and are given various luxuries. A Soviet soldier – to
pick just one item – lives mostly on black bread and soup, and
perhaps gets white bread three or four times a year. His life is
pared down to the bone, without pampering.

John Gunther
Inside Russia Today

In a recent book[1], Elizabeth David argues that it is probably the
tastelessness of modern commercial bread that leads to over-
consumption of so much sugary confection. Nobody could
seriously claim that the sliced stuff on offer in supermarkets and
chain bakeries gives much in the way of culinary delight. It looks,
feels and tastes like 'filler' no matter what the manufacturers and
their tame nutritionists say. As with that other staple, beer, there is
one safe and simple rule. If it has to be advertised it is probably
unfit for human consumption. The home-made product is the one
that disappears first at every church fête or bazaar. The nutty,
chewy, crusty loaf is almost a meal in itself and the breakfaster who
is 'happy with just a piece of toast' is always happier if it comes
from such a source.

Decent bread is, however, often hard to come by and is rarely
available in supermarkets. Most towns have at least one independent baker who sells a genuine loaf, but many of them have had
their standards depressed by the ubiquity of the plastic product
which now sets the pace. There are some very strange animals
masquerading under the 'brown', 'wholemeal', 'wheatmeal',
'granary' and 'wholewheat' labels. As in the so called 'hot bread'

kitchens the labels are a desperate attempt to inject some sort of culinary respectability into a plastic corpse. Wholefood shops usually market a wholemeal bread made virtuously from ingredients of the highest pedigree. Often, however, its unimaginative baking to a woody texture gives easy ammunition to the junk food anti-health propagandists.

Best results, as ever, come from the home-baked product. The prospect need hold no terrors. Making bread is neither difficult nor very time-consuming since most of the preparation consists of leaving the dough alone to do its own thing. When we add that it is child's play, this is a literal truth. A bored child will respond with enthusiasm to the prospect of helping with the breadmaking and even more to the magic of a very own and personal, private loaf.

Something as fundamental as bread inevitably rouses powerful emotions. Like sugar, it attracts more than its fair share of symbolism and myth. Its colour and integrity are matters for passionate debate. At one end of the spectrum the wholefooders will eat nothing but the whole, unsprayed grain, while others see social status and cultural superiority in the light, white and well-raised product. None of them has much to say about taste and this is surely the heart of the matter. Since bread is a staple and nourishing food, the object of the exercise ought to be to make it attractive so that it will be eaten. The dieticians also make one wonder whether they actually enjoy eating food as opposed to consuming nutrients, for whenever bread is discussed in a health context, they knee-jerk exclusively towards wholemeal. White bad, wholemeal good seems to be the whole of it.

In fact the world of bread is vast. Not all white bread is bad and wholewheat is often dull. Between them stretches a range of tasty possibilities, some of which we include below. Additions of rye flour and flake, oat and oatmeals, wheats in their varieties, barley, soya products, seeds and nuts add taste, variety and, dare one say, joy to this world. They also have the nutrients in sufficient profusion to satisfy the list-makers.

Wholewheat breads are of course marvellous if properly prepared from strong flours carefully baked. Commercially made from low-gluten weak flours which have been agitated to lightness by the Chorley Wood process they seem to be little more than a Browning version of supermarket plastic white. As far as taste goes they issue nothingness. Dull and tasteless as most white breads are, children reared on a combination of wholewheat and sermon-ising are likely to find the attractions of the white supermarket sponge irresistible. We therefore include a recipe for a (pseudo) white loaf to meet this curiosity.

Sharp-eyed readers who have made bread before might wonder at the absence in our recipes of the sugary starter which is usually puddled into the middle of the flour. Most other recipes do include it but in practice it is quite unnecessary. We have made bread and other yeasted products for years without it and it was reassuring for us to read that Elizabeth David has not only had the same experience but considers that prolonged direct contact between sugar and yeast cells can destroy the latter.

We will be returning to the subject again in Chapter 9 where we find that yeast is one of the useful keys to a lively and healthy diet, but as many cooks approach it with some nervousness we should add a word or two of explanation. Many first-timers fail because the nature of yeast is not fully explained. Yeast is a plant, a fungus in fact, which functions by turning sugars and other carbohydrates into alcohol, and this is accompanied by the emission of carbon dioxide gas. It is this gas which aerates the dough to give a light texture. Like human beings, it works most efficiently at blood heat. As the temperature drops it slows down. If the temperature gets much above blood heat it dies. So instructions about times of rising and proving are not helpful unless related to surrounding temperatures. Covered with a damp tea towel, dough can rise quickly in a hot kitchen or airing cupboard. If such conditions are not available the best thing is to leave it overnight for a slow rise which in fact often improves taste and texture. Rising can often be speeded also by using more yeast. This will leave its mark in flavour which some may find unwelcome. We tend to use less yeast than most orthodox recipes, but this is because we choose to give the dough lots of time.

The important thing is not to be disheartened by early failure, and to follow early success by a repeat performance as soon as possible. Once mastery of yeast is achieved there will be no end to the pleasure of cooking with it.

'Brown' bread
(sufficient for two 1kg (2lb) loaves and one 500g (1lb) loaf)

1kg (2lb) strong white flour
1kg (2lb) wholewheat flour
2–3 tsp salt
1 tsp dried yeast
a little milk

1 litre (1¾ pints) warm water
15g (½oz) bran
25g (1oz) wheat germ
    (optional)

Blend the flours and salt in a large warm mixing bowl and place in a warm spot near a cooker or in an airing cupboard. Put the yeast in a small jug or mug, add a little milk and ¼ pint warm water. Cover with a saucer and allow yeast to disperse (about 20 minutes). Make a well in the centre of the flour. Pour in the yeast mixture, stirring in a little flour. Gradually add the remaining 1½ pints warm water, all the time incorporating more flour. Continue kneading with your hands until the dough is no longer sticky. Add more water if it seems too dry and more flour if it seems too sticky. Make the dough into a ball in the centre of the bowl and cover with a damp tea towel. Allow to rise in a warm place until the dough has doubled its size (approximately 2 hours). If you leave the dough to rise overnight at room temperature, you can make fresh bread for breakfast.

Heat the oven to 220°C/425°F (gas mark 7). Oil two 1kg (2lb) loaf tins and one 500g (1lb) loaf tin. Place the bowl of dough on a firm surface. Knock down the dough and turn out on to a floured board. Knead vigorously for a minute or two. Cut off about one fifth of the dough and form into a loaf for the smaller tin, placing the folds at the bottom. Divide the remaining dough into two and form into loaves for the larger tins. Slash the top of each loaf three or four times with a sharp knife. Cover with a damp tea towel and leave to rise to almost twice its size in a warm place (this can take anything from 30 minutes to just over an hour). Bake for 20 minutes. Then turn the temperature down to 190°C/375°F (gas mark 5) and bake for a further 25 minutes. Remove from the tins. If cooked, the bread will sound hollow when you tap the bottom with your knuckle. If not, bake for a further ten minutes out of the tins. Cool on a wire rack. Hot bread is irresistible but, if you can, let it get cold. It will slice more easily and the flavour improves with cooling.

'White' bread

1 tsp dried yeast

8cl (3fl oz) skimmed milk
  made up to 30cl (½ pint)
  with warm water

1.25kg (2¾lb) strong white
  flour

40g (1½oz) wheat germ

25g (1oz) bran

7g (¼oz) soya flour

2 tsps salt

60cl (1 pint) warm water

'Dissolve' (i.e. disperse–it doesn't actually dissolve) the dried yeast
in the diluted milk for about 20 minutes. Reserve about 125g (4oz)
flour for kneading. In a warm mixing bowl blend the rest of the
flour, wheat germ, bran, soya flour and salt. Make a well in the
centre. Pour in the yeast mixture, gradually incorporating a little
flour. Add the remaining warm water slowly, stirring in the flour
until the dough is too stiff to use a spoon. Knead with hands until
the dough is no longer lumpy. It will be a little sticky at this stage.
Cover with a damp cloth and leave to rise in a warm place for about
2 hours or until the dough has doubled in size. Knock down and
turn out on to a floured board and with the remaining flour
re-knead. Divide into two 1kg (2lb) oiled bread tins. Slash the top
of each loaf three or four times with a sharp knife. Cover and leave
to rise in a warm place for a further 45 minutes or until the dough
has risen just above the top of the tins.

Bake in a preheated oven at 220°C/425°F (gas mark 7) for 40
minutes. Remove from the tins and test by tapping the bottom of
the loaves with your knuckle. If the sound is hollow, the bread is
cooked. If not, bake out of the tins for a further 10 to 15 minutes.
Cool on a wire rack before slicing.

Maslin bread

In an earlier chapter we noted Sir Jack Drummond's analysis of
the fifteenth-century peasant's daily (and dentally healthy) fare.
This included two pounds of maslin bread. The word 'maslin'
refers to a mixed sowing of grains. The usual mixture was wheat
and rye and, since the bread deriving from it formed a substantial
part of the diet, we felt that it might be of dental value in modern
mouths. It proved a rewarding venture, for the bread itself, though

unprepossessing in appearance, is remarkably tasty. Moreover it has a shelf-life of over a week. Here's a modern version.

500g (1lb) wholewheat flour	approx. 60cl (1 pint) warm water
500g (1lb) rye flour or medium ground oatmeal	8cl (3fl oz) skimmed milk
250g (8oz) strong white flour	1 tsp dried yeast
2 tsp salt	

Dissolve the yeast in the milk and 15cl (¼ pint) warm water and leave for about 20 minutes. Blend all the dry ingredients in a warmed mixing bowl. Make a well in the centre and pour in the dissolved yeast mixture. Gradually incorporate the flour and the rest of the water, stirring with a wooden spoon. When the mixture becomes too stiff, abandon the spoon and knead the dough throughly by hand. When no longer lumpy, shape the dough into a ball, cut a deep cross in the top with a sharp knife and cover with a damp cloth. Leave to rise in a warm place until the dough has doubled in size (this may take 2 hours, but it is better left overnight).

Knock the dough down and turn out on to a floured board. Re-knead, then divide and shape into three loaves for oiled 500g (1lb) bread tins. Prick the top several times with a fork. Cover with a damp cloth and allow to rise for a further 30 minutes to 1 hour.

Preheat the oven to 220°C/425°F (gas mark 7) and bake for 40 minutes. Remove from the tins and bake for a further 10 minutes. Allow to become quite cold on a wire rack before cutting.

Oatmeal sesame plait

500g (1lb) strong white flour	1 tsp salt
250g (8oz) medium oatmeal	½ tsp dried yeast
50g (2oz) wheatgerm	45cl (¾ pint) warm water
40g (1½oz) soya flour	Skimmed milk
50g (2oz) rolled oats (or porridge oats)	Sesame seeds

Blend the flour, oatmeal, wheatgerm, soya flour, oats and salt.

Make a well in the centre. Put the dried yeast in the well with 15cl (¼ pint) warm water and lightly cover with a sprinkling of flour. After ten minutes the yeast will have dispersed. Add the remaining warm water, gradually incorporating the flour until it can be kneaded by hand. A little extra water may be necessary if the dough seems too dry. Knead vigorously until the dough is no longer sticky and forms a ball in the centre of the bowl. Cover and leave to rise in a warm place until the dough has doubled in size (this may take 2 hours in a very warm environment, but it is better left overnight).

Re-knead the risen dough and divide into three equal parts on a floured board. Hand roll each piece into a long sausage. Lay them side by side. Press together at one end and plait loosely. Lift carefully on to an oiled baking sheet. Brush the top with skimmed milk and sprinkle generously with sesame seeds. Cover and leave to rise in a warm place for a further hour or until the plait is well risen. Bake at 220°C/425°F (gas mark 7) for 25–35 minutes. Cool on a wire rack.

Rye and Kibbled Wheat Loaf
(makes two 500g (1lb) loaves)

250g (8oz) strong white flour	½ tsp yeast
250g (8oz) wholewheat flour	1 tsp salt
250g (8oz) rye flour	45cl (¾ pint) warm water
50g (2oz) kibbled wheat	
50g (2oz) wheat germ	*Glaze* (optional)
25g (1oz) soya flour	Potato flour and water

Blend all dry ingredients except the yeast. Make a well in the centre. Put the yeast in the well and add 15cl (¼ pint) warm water; lightly sprinkle with flour. Leave for the yeast to disperse; this will take ten minutes. Add the remaining water, stirring in the flour until the dough can be kneaded by hand. Knead vigorously until the dough is no longer sticky and forms a ball in the centre of the bowl. It may be necessary to add extra water if the dough is too dry. Cover and leave to rise in a warm place overnight.

Re-knead the risen dough. Divide into two and shape into loaves. Place in two well oiled baking tins. Prick the top with a fork. Glaze with a wash made from 1 tsp potato flour and a little water. Cover and leave to rise in a warm place for about one hour. Bake at 220°C/425°F (gas mark 7) for 40–50 minutes. Turn out of the tins and test by rapping the bottom with the knuckles. If it sounds hollow it is cooked, if not return to the oven for a further ten minutes. Cool on a wire rack before slicing.

FREEZING BREAD

Bread freezes very well. Once it is hand hot, fast freeze it (unwrapped) on a wire rack, then wrap in polythene bags. We are aware that this is contrary to usual freezer practice[2] but it appears to prevent staling. It will keep for about four months. When defrosting, the crust may tend to break. To avoid this, cool in a moist atmosphere after baking.

SPREADS, JAMS, etc.

The question of which fatty spread, butter or polyunsaturated margarine, goes on to the bread, whether toasted or fresh, holds no particular *dental* interest, however much it touches on general medical concerns. It is the second layer we have to think about, as this usually consists of some kind of sugar concentrate, either honey or what is genteely termed a 'preserve'. Honey has a wholesome image and claims are made for its magical and sexual powers. It remains, however, a sugar concentrate and there is nothing in its so-called 'naturalness' to affect its tooth-destroying powers. Jams, on the other hand, are really mummified fruits, which have been drowned in sugar to make a death trap for the spoiling bacteria. The word 'preserve' is inappropriate in this context. If there were some way of removing the sugar, leaving the fruit more or less restored, then the word would have some justification. After all when the preserving low temperature is withdrawn from deep-frozen fruit they just about reconstitute, give or take a little limpness. In jams, however, the fruits' delicacies of flavour are drowned for ever in a sugary cocoon and, even though their acidity converts much of the sucrose into invert sugar (i.e. fructose and glucose), the result is custom-built plaque food.

Here we can therefore make the first of our borrowings from the world of diabetics, for whom sugar is a menace to general health as well. To sweeten food they use two alternative substances, sorbitol and saccharin. Sorbitol is about half as sweet as sucrose but behaves fairly similarly in cooking. It is more expensive than sugar and for some people, if taken in excess of 50 grams (two ounces) a day, can have a slight laxative effect. It is often backed up by saccharin, a substance of no food value at all but which is 200 times sweeter than sugar. Unfortunately it has a slightly unpleasant after-taste for some people. Manufacturers must therefore juggle the quantities of these two ingredients to get the best results. As if in compensation, they tend to create products which are far sweeter than anything on the supermarket shelf or in the standard cookery books. This limits the value of commercial diabetic products for readers who are aiming at sane levels of sweetness. These products will be discussed further in the book.

Back at the breakfast table, we find that the sorbitol-sweetened marmalades produced by Frank Cooper and Boots are of high quality. In jams, alas, the sweetness if often overdone and their use should be confined to rare occasions like parties. Indeed even the marmalades should be sparingly used. Sorbitol may not be the most powerful of plaque fuels but, given an excess of it, natural selection might well ensure that it soon was. The keeping quality is not as good as that of the sugared varieties and, once opened, the jars should be kept in the refrigerator. We have experimented with a range of home-made spreads and some of these will be appropriate for breakfast. Recipes for these are found in Chapter 9.

Drinks to accompany breakfast food present no problems provided sugar is not used to sweeten them. Fruit juice should be unsweetened, whether it is bottled, tinned or frozen. Anything called a 'fruit drink' or a 'Vitamin C drink' should be avoided like the plague it is. The unsugared exceptions are listed in Appendix 1. If you are a Vitamin C fan, then add fresh fruit or any of the fruits tinned in natural or apple juice (listed in Appendix 1).

One point remains; tea has a high fluoride content, and it has been authoritatively suggested that the British tea habit has kept the dental disease level lower in Britain than in other countries where less sugar is consumed but also less tea. The difference is not, however, all that great and heavy tea-drinking will give little immunity to the sugar addict.

9 Half-time scores
Snacking safely

My reason is my friend, yours is a cheat;
Hunger calls out, my reason bids me eat;
Perversely, yours your appetite does mock:
This asks for food, that answers, 'What's o'clock?'

John Wilmot, Earl of Rochester

NORA:
No, no, don't be frightened; you weren't to know that
Torvald had forbidden them. The thing is, he's afraid I shall
spoil my teeth with them. But pooh–just this once! That's
right isn't it Dr Rank? Here! (*She pops a macaroon into his
mouth.*) And now you, Kristina. And I'll have one as well–
just a little one. Or two at the most.

Henrik Ibsen
A Doll's House (1879)

Ideally, given a sensible spread of main meals, there should be no
nutritional need for any snack in between. In the real world,
however, we have rituals like elevenses and tea-breaks, which are
enjoyable and important. They shape a yawning day, offer focal
points for entertainment and social intercourse and provide rest
from demanding work. It is also true that for certain people, those
at lean weight, breast-feeding mothers, adolescents and highly
active children, there can be a genuine hunger problem. Moreover
the very notion of a clear distinction may already be an outdated
middle-class concept. This concept was always pregnant with
moral overtones. On the one hand were sensible, balanced meals
known as a good thing. On the other hand there were wicked
snacks which good children who had eaten up their porridge ought
not to want. It is an attitude which has enabled the advertisers to

market high fat, high sugar products under the 'Naughty but Nice' label.

Evidence is growing that many children in particular are not eating a midday meal at all, and a significant number have snacks instead of an evening meal.[1] Many indeed see meals as an unwelcome interruption to steady snacking. All of this makes a dog's breakfast of the simple-minded dental scientists' view, 'sugar at meals OK, snacks bad'. Snacking is clearly here to stay and perhaps expand its role. Our concern must now be to compete successfully with the junk salesmen and enrich ourselves with food that is tasty, nourishing and harmless.

It is in this aspect of our eating above all that the sugar culture claims the field. Our old friends 'zip' and 'zest' are the genies produced from the sugar bag whenever our old enemies 'faintness' and 'lassitude' threaten to strike us down between meals. In the fantasy world of the sugarmen's public relations officers each of us is stalked by the phantom of between-meal hunger when muscle and liver are drained of energy stock, leaving blood dangerously thinned and brain on the blink. As we sag and our limp bodies threaten to pile up on factory and classroom floor, and fingers tremble weakly towards the 999 call, suddenly in to the scene leaps 'Supersnack' the sugary redeemer that will supercharge the faltering engine and revive the sapped spirit. For at this level merely 'liking' the confections is too mundane and guilt-ridden a motive for approaching them. We have to see them as 'nutritional first-aid'[2] an expedient not available to our uncivilized Saxon ancestors.

Such exhaustion as envisaged above may describe the condition of an athlete down to lean weight and *in extremis*. *Our* daily hungers are of the milder sort, and our average fatigues simply require rest. In the course of the ordinary day our energy reserves are rarely exhausted and, alas, are often only too prominent. To meet everyday deficits we need food of a balanced kind, not high energy confections, addiction to which will merely make us fat.

So maybe our first thoughts should be of the simple expedients, like bread and scrape, a cream cracker or two, a piece of fruit or a chunk of mousetrap. Time was when that was the obvious answer, but nowadays the unremitting din of commercial advice dulls out such common sense. There is, however, a wide variety of unsugary snack foods, and no junk-fooder with a glimmer of dental ambition need feel unhelped.

At the centre of this scene is our good friend the potato, or rather its crispy offspring, which has long been with us in its most simple form. Now its new liaisons with the cereals have opened up a whole universe of snack variety. Potato in the raw has about 0.5% of free sugar, not a dose to get any dentist's adrenalin on the surge. Manufacturers anyway seem to prefer a potato with something less than 0.3%, making the plain crisp, even without its saliva-stimulating salt, probably the safest snack food this side of the Dark Ages. The fat and salt content is, however, high so, quite literally, don't make a meal of them!

The sugar culture, has another card to play. Mere salt on crisps is not enough. We can now have breakfast, dinner, and tea channelled through the crisp bag in the form of bacon and egg, chicken, beef, ham, cheese and many other flavours. These flavours are chemical spices manufactured by 'flavour houses' and their carrier is sugar, sprayed on to the cooked surface, where they attach in such quantity that up to 4% of bag weight can be sugar. Comprehensive evidence is not easy to come by but, on the material supplied to us by one firm (Walker), it looks as if the least sugared are the plain, cheese, cheese and onion, beef and onion, bacon and smoky bacon flavours, while the spray-on sugar can approach 4% when the flavour is tomato or salt and vinegar. Given that the salt and flavour level is always high, we can expect a rapid saliva flow when they enter the mouth and, as we saw in an earlier chapter, copious saliva is the great mouth trouble shooter. Indeed a small experiment quoted by Walkers showed that the acidity on the tongue after eating a packet of crisps, whether the flavour was 'acidic' (salt and vinegar) or 'neutral' (cheese) fell in each case.

We can therefore conclude that the crisp of whatever flavour is well up in the dental safety charts. If, however, the words 'sugar', 'dextrose', 'lactose' or 'glucose' appear separately on the ingredient panel of the packet, you should look elsewhere. One firm (which had better remain nameless) records 'glucose solids' as the second ingredient even in obviously savoury varieties! The phenomenon is not widespread but the habit of inspecting the ingredients panel is worth some cultivation. Those panels checked by us are listed in Appendix 1.

If the world of the crisp has become more exciting and diversified, this cannot be said for bread. Its only value is its low sugar content. But since research shows that in the USA and USSR unsweetened bread is becoming a rarity, then perhaps, with such newer horrors on the horizon, we should be grateful for what we have. Meanwhile, as we have stated earlier, we can still do our own thing and the freezer makes this more feasible. Otherwise we can at least track down an independent baker who makes something that is a pleasure to eat.

There are, however, new diversities in those dehydrated products known as crispbreads. Their fundamental appeal is magical rather than scientific, even when so-called 'starch-reduction' is a feature. The magic lies in the hope that the thinness of the slice will somehow infect its overweight consumer in the same way that thin chocolate-covered mints are expected to spirit away the consequences of a four-course dinner. The crispbreads are, however, both pleasant to eat and safe vehicles for spreads in half-time sit-down rituals. Few of them are sugared and those that are are nearly always the ones with a high bran content. (Bran, whatever other physiological liberations it stimulates, certainly seems to open the sugar floodgates. Kellogg's 'All-Bran', for instance, manages a total free sugar percentage of 15.4.)[3] Without crispbreads, however, bran roughage is still easily available elsewhere in the diet. The safe ones are listed in Appendix 1.

Pastes, spreads and butters constitute something of a minor dental minefield. Having rejected the obvious sugary items like jam and honey, we have to inspect the alternative pastes, spreads and butters for the sugar contaminant. Peanut butter, absurdly enough, is rarely available commercially without sugar. Its home preparation is fortunately a simple matter. Meat and fish pastes occasionally feature sugar low down on the panel and it may also come through the spice, so we list the safe ones in Appendix 1. Cheese spread is rarely sugared and then only when it incorporates some flavour like sweet pickle. We offer a few pointers for those who both prefer and find time to do their own versions. Further variety is available at the delicatessen counters of the supermarkets and in the standard recipe books, which should be consulted for spreads such as hummus, taramasalata and cheese dips.

Peanut butter

500g (1lb) unsalted peanuts   salt to taste
4 tbsp cooking oil

Preheat oven to 190°C/375°F (gas mark 5). Spread the peanuts evenly on a baking sheet and roast lightly for 20 minutes turning after ten minutes to ensure even roasting. Cool. Depending on taste, grind finely or roughly in liquidizer. When the required consistency is reached, add salt and oil and continue grinding until blended. Store in a screw-top jar. Keeps indefinitely.

Cashew nut butter

500g (1lb) cashew nuts   salt to taste (very little is
4 tbsp cooking oil   needed)

Grind the nuts in a liquidizer to fine consistency. Add oil and salt and continue grinding until blended. Store in a screw-top jar. Keeps indefinitely.

RAW DEALS

The advocates of raw food come in different shapes and sizes. The hard-liners would have it that the discovery of cooking was the beginning of the nutritional end, that it amounts to a hazardous tampering with the bounties of a benevolent nature. Even those who look a bit further than the Psalms of David for their nutritional information may remain suspicious of domestic technology and find rectitude only in a preference for raw materials. When vitamins became the fashion and the discovery was made that some of them could be destroyed by heat this fad seemed to be underpinned by solid science.

It is of course true that certain vitamins (B1, C, and folic acid) are heat sensitive, though careful cooking can minimize losses. It is equally true that the uncooked potato and the unreconstituted pulse are nobody's treats. We need heat treatment also to destroy the spoiling bacteria that feature in our stomach upsets. Alas, for our purposes, neither nature nor nutrition offer those simple and symmetrical guidelines that so beguile the healthfooder.

The faddist's zeal together with the dietician's dreary handout literature have conspired to give raw food the forbidding image of something that is good for us, and many a hungry child's burgeoning summer days have been blighted by unimaginative salad, when he/she yearns for *real* food.

We do much better if we stick to the pleasure principle, because raw food can be a delight. If we have to do good we do it by stealth, and perhaps even without effort. At the teething stage, as we have seen, a slightly chilled raw vegetable will soothe an aching gum, and a good image is there for the taking. Later, during meal preparation, our commonplace vegetable turns into stolen fruit as slices are carried from the sink in triumph. More stylish preparations can accompany the sit-down ritual. For instance, thin carrot strips, celery curls, cauliflower florets, broad beans and so on can be eaten straight. Cucumber and the hard root vegetables like swede, parsnip and kohlrabi are often improved by a sprinkling of lemon or orange juice. A trick with tomatoes is to cut them in half in a zig-zag pattern with a sharp knife.

Citrus fruits with easily detachable skins can be turned into 'water lilies'. Placing the stalk end upwards, eight incisions are made lengthways, through the skin only, at regular intervals and continued almost to base. The skin segments are peeled back from the fruit but not detached at the base. The fruit segments are separated, after which the whole fruit is 'closed' and brought to the table where it opens magically into a lily.

Sweet safeties

The conventional dental wisdom about diet stresses the importance of avoiding sugar between meals. Occasionally there is a rider that consumption in general should be reduced but this part is seldom offered with any enthusiasm. Studies show that the frequent presence of sugar in the mouth is a more important factor than mere quantity. This neat conclusion is tossed to the public in the expectation that they will immediately alter their ways. That they don't is not seen as showing the naivety of the advisers, but rather the fecklessness of the advised. The further conclusion is then made, in the profession's internal literature at least, that dietary control is unrealistic.

This advice is, also, sometimes open to misinterpretation. It has become in some families a licence to gorge large amounts of sweet confections at mealtimes, to compensate for interim abstention. There is evidence[4] to show that this practice can be dangerous, leading to unhealthy surges of blood tri-glycerides (fats) and possibly obesity and coronary heart disease.

Another consequence may be the phenomenon known technically as 'reactive hypoglycaemia'. This means low blood sugar caused by high sugar intake. The body has a sensitive mechanism to keep our blood sugar level constant. After a meal has been digested, sugars and other products flow into the bloodstream from the intestine and blood sugar rises. This rise stimulates the pancreas to produce insulin which in turn causes the liver to take the excess sugar out of the circulation. Conversely, in starvation the level is maintained because another gland causes the liver to release stored sugar. The system works well when the diet is composed of natural unconcentrated foods because the sugar in them is released slowly. Heavily sugared foods, however, cause a rapid squirting of sugar into the bloodstream which overwhelms the pancreas which then over-reacts. Too much insulin is produced so that the overall effect of taking a high sugar dose is to cause a *drop* in blood sugar and a craving for food. Domestically this shows up as a three-thirty demand as children return from school shouting, 'I'M HUNGRY!'

It is suggested by some that constant overstimulation of the pancreas in this way may be a strong factor in the development of diabetes. This may well be controversial, but what is not in doubt is that the dentist's advice to confine sugar eating to mealtimes may be counterproductive in provoking a craving for more of the same shortly after a meal.

However beguilingly simple the sugar 'frequency' argument seems, we need a broader base for our strategy. This is provided by history. We know from skull studies that towards the end of the Middle Ages, tooth decay was exclusive to the rich. As sugar grew cheaper, decay began to reach the middle classes, until in the middle of the last century when sugar tax was abolished it was finally democratized. The most frequent cause for the rejection of Boer War volunteers was bad teeth.[5] By the turn of the century sugar reached its high noon of destructiveness and this peak persisted well into the thirties when there was a sudden reversal associated with the outbreak of the Second World War. Hitherto there had been an uninterrupted growth in the accessibility of sugar, accompanied by an increase in tooth decay. Now to help refine the hypothesis that the two were causally connected we were able to look at what happened when the supply was reduced.

One study[6] of this was made in the two towns which straddle the Tyne estuary, North and South Shields. The dietary change caused by the war was similar for both populations. At this time South Shields' water supply had a natural fluoride content of 1.4 parts per million, while North Shields had hardly any. It is now well known that the decay experience of South Shields children was less than half that of their peer groups over the river. What is less well known is that in the period 1943 to 1949 the North Shields teeth improved to the pre-war level of the South Shields teeth, and this must be attributable to the wartime diet. At 67lbs of sugar per annum (i.e. about two-thirds of current consumption) this can by no means be described as low sugar.

Studies of wartime populations in Japan, Norway, and the Channel Islands showed even more dramatic improvements in dental health from reductions in annual sugar intake. A serious prediction has in fact been made that if sugar consumption can fall to below 10kg a year (22lbs)[7] then, with a little help from fluoride, dental disease can be eliminated.

This book aims to keep us well within this limit by a number of simple stratagems and a concurrent refusal to let in bag sugar at all.

Sweetness need not be and is not absent here. Two general points must be made. First the sweet snack cannot be routine. It should take its place in rotation with the savoury and raw alternatives. Secondly, current consumption of sugar in this country runs at just over four times the 10kg limit we are aiming for and we are therefore trying to reduce the sugar content in our recipes by just over 75%. This reduction can be made by altering the traditional cookery book ratio of sugar to cereal from 1:1 to 1:4, with actual gains in taste. The sugar we use is of course not from bags but from dried fruits, thus ensuring a further variety in taste. The sugar content of these fruits averages about 60% and to show how this makes a superior product we have now to look at two further stratagems that make it possible.

The original purpose of spices may have been to smother the smell of unfresh materials, particularly meats and fish, but they had medicinal uses as well as the straightforward one of bringing out the flavours of blander ingredients. There is some evidence that spices have long featured in the eating habits of Britain, but it was the Romans who introduced varieties such as pepper, ginger, saffron, coriander and mustard. Sugar as such was not used but the widespread use of honey and imported dried fruits may have been responsible for the significant increase in dental disease which skull studies[8] show during the Roman-British period. It was, however, never remotely in the same class as our modern epidemic sort.

In the Anglo-Saxon period there was a lull in the importation of both spices and the dried fruits. Honey was still used to lift the blander cereal flavours but consumption of its sugars cannot have been anywhere near heroic twentieth century standards and the skulls show a return to Bronze Age tooth integrity. The Norman conquest brought a revival of spice usage and throughout the late Middle Ages it loomed large in trade. Gradually sugar began to penetrate the market, at first only for the very rich. There remained a widespread passion for strong aromatic flavours which persisted even in those rich country households where the question of the unfreshness of meat could not have arisen.

During the eighteenth century the amount of spice in recipes declined and the level of sugar began to climb. In the following centuries the trend continued and its apotheosis might well be seen in that peculiarly English phenomenon, traditional victoria sandwich cake, the staple item of every church fête and sale-of-work. Given that its traditional sugar-cereal ratio is 1:1 there is not much room for any other flavour and none is usually put in. The taste is unmodified sweetness and nothing else. The displacement had now become total.

So we have to reverse the process somewhat. If we bring back the spices we can reduce the sugar and this enables us to present a variety of cakes well within our low sugar to cereal ratio.

The historical slither from spice to sponge carried in its wake a decline in the use of yeast. If aeration could be achieved by a combination of egg foam, baking powder, sugar granules and elbow grease, the nervous cook could be safe from yeast's mysterious power. To the uninitiated there is something about the faint tremble of rising dough that suggests that an uncontrollable force is at work, that this strange microscopic entity might easily engulf our culinary ambitions with its apparent wantonness. Yet, quite simply, yeast is a great labour-saving device. Millions of tiny cells, given a little warmth and moisture, will lift many times their own weight in dough by their fermenting power. There is nothing in the sponge cake to rival this texture, and the absence of sugar from the scene leaves the ingredient list wide open to our panoply of spices and flavours.

At the commercial level there is not much on offer. The sweet, candy, or other confection must rank, no matter which studies inform our judgement, as the deadliest enemy. Within this generality there is some evidence that enables us to rank the sweets in order of danger. The discovery, for instance, that in a study of human subjects, sucking an iced lolly caused no increase in the plaque acidity should perhaps make us think twice about our wide condemnation. But having thought twice about it, there seems no reason to revise our view that every time heavy sugar concentration enters the scene, addiction is being sustained no matter how cheerful laboratory results look. In this case, the most likely causes of the low acidity may be the lowered temperature of the plaque and the high saliva flow which the ice provokes. This can be achieved just as easily in the home if freezing facilities are available. Unsweetened fruit juices, apple, orange, pineapple or grapefruit, are simply poured into plastic moulds and frozen, thus giving the benefit of saliva stimulation without added sugar.

Another oddity, as we saw in Chapter 3 is liquorice. This is available in many shapes and even colours. Though made with sugar it does not seem to increase plaque acidity to danger level. Given that, unlike the lolly, no safer sugar-free substitute is available, we make this the single exception to the no-sugar rule, though in our own family its consumption is something like a monthly or festival event. Our approach to diabetic foods is similar. Although in general chocolate seems to be the least dangerous of the orthodox sweets, we have preferred to stick to those brands sweetened without sugar and labelled 'diabetic'. The sweeteners are either sorbitol (a natural sweetener found in rowanberries) or fructose (see Chapter 3). The other diabetic sweets of the boiled type are high in acid and spend a long time in the mouth. This must take them down in the safety charts with their sugared fellows.

A possible new growth area of sweet significance is provided by the macrobiotic world which will have no truck with sugar. The sweets they sell through their own and some wholefood outlets are sweetened with rice malt, one of the least dangerous of the sugars. Like liquorice and diabetic sweets they should, however, be seen as treats and not part of a daily cornucopia to be unloaded at the school gates at four in the afternoon.

Most chewing gum is sugared. Two brands, however, 'Orbit' and 'Trident', are sweetened with other agents (mannitol, sorbitol, saccharin), and can be taken as dentally safe.

On the drinks side, it is clearly established that sugared beverages cause rapid increases in plaque acidity. We have listed the low-calorie beverages which are sugar free in Appendix 1 and only note here that not all low-calorie drinks are sugar-free. Waitrose low-cal tonic water, for instance, is sweetened with fructose. Safe drinks are milk, unsugared tea, coffee (instant or ground) and cocoa (a mixture of cocoa and diabetic drinking chocolate produces a sweet but not over-sweet drink–the diabetic stuff alone is painfully sweet). There are, of course, the natural fruit juices, frozen, canned or bottled, as well as 'Oxo', 'Bovril' or 'Marmite'. The baddies are the ordinary lemonades, rosehip and blackcurrant syrups at any dilution, no matter how rich in vitamin C, glucose drinks, however much they are alleged to fortify the convalescent and sick, milk shake mixtures, bottled coffee essence, fruit squashes, ginger beer, colas, tonic waters (unless unsugared) and malted bedtime drinks.

Those readers with very young children should have no problems here if the previous chapters have been followed, but there will be others who have hitched on to the book without this advantage and who have difficulty in reversing long established patterns. Those with schoolchildren may find themselves without support in the community. Schools, in particular, with their tuckshops and PTA events where sweets play a prominent part, constitute a powerful authorizing force which can isolate the enterprising parent. There are tricks which can be employed in such circumstances but none of them replaces the total commitment that alone can win the battle against a complacent environment. Parents must simply decide that they are masters of their childrens' fate at least until teenage years; that their health is a matter to be decided not by other people's social habits, but by their own resolve. It is in fact surprisingly easy to gain the acquiescence of others provided a firm enough stance is adopted. One of our local shopkeepers, being kindly disposed to our children, offered sweets and was told with gratitude and pleasure at the kind thought, 'They don't eat sweets'. The result was simply that they now have gifts of fruit. Grandparents, for whom the sweet gift is made easy by the strategy of shopkeepers, may need preliminary diplomacy. This is often a matter of listing alternatives such as fruit, crisps, crayons, colouring books, balloons, according to current interest. It follows that the anti-sugar stance should be a whole family matter and not a penance imposed upon selected members.

There is a kind of stage fright in attempting what seems at first to be a highly unusual departure. But once the first leap is made it is quite surprising how much hidden support can be flushed out.

Most communities harbour a dormant feeling that all is not well with the sugar scene but this is accompanied by an all too prevalent sense of helplessness. The former should be worked on and a little courage should be applied to combat the latter. Children of five and over who are subject to pressure from their peer group at school are nevertheless, for another six or seven years at least, still susceptible to a parent's calm appeal to their common sense. It may well be that more immediate compensation than future dental health may have to be injected into the argument. Bribery, though preferably a last resort, needs to be carefully planned for the totally hooked. Here a target period of about three months of total abstinence should be rewarded by a promise of a really substantial gift at the end of the period. This will have to be shown to have been financed out of the daily saving on sweet non-purchase, as well as perhaps a 'with profits' bonus from the parent. Usually the end of the period is marked by such a recovery of the formerly jaded palate that previous levels of sweetness intake seem intolerable and the battle needs no more energy input from parents.

Chocolate cake

75g (2½oz) dates
4 tbsp sunflower oil
30cl (½ pint) skimmed milk
1 tsp dried yeast
1 tbsp skimmed milk powder
1 egg
2½ tbsp cocoa powder
50g (2oz) wholewheat flour

125g (4oz) strong white flour
50g (2oz) rice flour
15g (½oz) soya flour
25g (1oz) desiccated coconut
50g (2oz) ground almonds
25g (1oz) currants
1 tsp grated mixed rind
   (orange, lemon, grapefruit)

In a saucepan warm together the dates, sunflower oil and milk. Liquidize, cool and when no more than blood heat add the yeast. Allow to disperse for ten minutes before adding skimmed milk powder, egg and cocoa powder. Blend and set aside.

Mix all the remaining ingredients. Stir in the yeast mixture thoroughly. Cover and leave in a warm place to rise for about 2 hours. Tip the mixture into a greased, lined cake tin, approx. 20cm (8 inches) in diameter. Cover and allow to rise for a further 45 minutes. Bake at 190°C/375°F (gas mark 5) for 40 minutes. Test with a knitting needle. If it comes out clean turn the cake onto a wire rack and cool before decorating (see page 113 for cocoa cashew spread).

Fresh fruit cake

2 cups wholewheat flour
1 cup rolled oats (or porridge
  oats)
½ cup wheat flakes
½ cup barley flakes
25g (1oz) soya flour
1 cup desiccated coconut
1 cup mixed broken nuts
1 cup raisins (or mixed dried
  fruit)
1 cup grated carrot

3 cups crushed fruit (pulp and
  juice) e.g. apples, bananas,
  pears, strawberries or
  apricots
15cl (¼ pint) sunflower oil
1 tbsp each sunflower and
  sesame seeds
1 tsp vanilla essence
1 tsp mixed spice
¼ tsp cumin
grated orange and lemon rind

The size of this cake depends on the size of cup. Be sure to use the same cup for measuring all ingredients at any one time.

In a large bowl mix all the ingredients into a soft dough. Press the mixture into an oiled bread tin. Bake at 180°C/350°F (gas mark 4) for 40 to 50 minutes when the top should be golden brown. Leave to cool in the tin for 30 minutes then turn out onto a wire rack and allow to become quite cold before cutting.

The cake will stay moist for about ten days if kept in an airtight tin. It freezes well and makes a good standby cake.

Yoghurt Cake

90g (3oz) sultanas
6 tbsp sunflower oil
17.5cl (6fl oz) yoghurt
1 tsp dried yeast
1 egg
juice and grated rind of 1
  orange
grated rind ½ lemon
125g (4oz) strong white flour

125g (4 oz) wholewheat flour
2 tbsp wheat germ
1 tbsp soya flour
50g (2oz) ground almonds
50g (2oz) desiccated coconut
1 tsp mixed spice
½ tsp ground cumin
whole almonds (to decorate)

In a saucepan warm the sultanas gently with the oil and yoghurt. Liquidize and cool to blood heat. Add the yeast and allow to disperse for 10 minutes or so. Blend in the egg and add the juice and orange and lemon rind.

Mix all the remaining ingredients except for the whole almonds. Add the yeast mixture and blend thoroughly. Cover and allow to rise in a warm place for 2 hours (or overnight if possible). Tip into an oiled, lined round cake tin (approx. 19cm×5cm). Decorate with whole almonds and cover with a circle of greaseproof paper. Place a tea towel over and leave to rise in a warm place for a further hour.

Bake for 45–50 minutes at 190°C/375°F (gas mark 5). Test with a knitting needle. If it comes out clean, turn the cake out onto a wire rack and cool. (If the needle is not clean continue baking for a few minutes more.)

Store in an airtight container.

Excellent for freezing or for use in trifles or puddings which require toasted cake crumbs.

Yeasted spicy fruit cake

15cl (¼ pint) unsweetened orange juice
1 medium-sized eating apple (cored)
1 tsp dried yeast
1 egg
6 tbsp sunflower oil
1 tbsp brandy (optional)
125g (4oz) strong white flour
50g (2oz) rice flour
15g (½oz) soya flour (optional)

125g (4oz) wheat and barley flakes
25g (1oz) desiccated coconut
50g (2oz) ground almonds
50g (2oz) walnuts
50g (2oz) cashew nuts
50g (2oz) grated carrot
125g (4oz) mixed dried fruit
grated rind & juice of ½ orange and ½ lemon
1 tbsp mixed spice

Warm the orange juice to just below blood heat. Liquidize with the apple. Add the yeast and allow to disperse for ten minutes. Blend in egg, oil and brandy if used and set aside.

Mix all the remaining ingredients. Stir in the yeast mixture until thoroughly blended. Cover and allow to rise in a warm place for a couple of hours.

Tip the mixture into an oiled, lined cake tin (approx. 16cm

(6½in) in diameter). Cover with a circle of greaseproof paper. Put a tea towel over and let it rise for a further hour.

Bake at 200°C/400°F (gas mark 6) for 50 minutes. Reduce the heat to 180°C/350°F (gas mark 4) and continue baking for a further 20 minutes. Remove from the tin and cool on a wire rack. Store in an airtight container.

Fruit loaf

½ tsp dried yeast
4 tbsp skimmed milk
30cl (½ pint) warm water
500g (1lb) strong white flour
25g (1oz) bran
25g (1oz) wheat germ
15g (½oz) soya flour
3 tsp mixed spice

½ tsp salt
50g (2oz) each sultanas,
  raisins and currants
2 tsp lemon juice
grated rind of ½ orange and
  ½ lemon
4 tbsp sunflower oil
juice of ½ orange

Dissolve the dried yeast in milk and warm water. Blend the dry ingredients, the grated rinds and the oil. Make a well in the centre and pour in the yeast mixture and lemon juice. Gradually incorporate the flour to form a slightly sticky ball of dough. Cover with a damp cloth and leave to double in size in a warm place. Knock the dough down and turn on to a floured board. Knead, then shape into a loaf and put into an oiled 1kg (2lb) bread tin. Prick with a fork, cover and allow to rise for a further 45 minutes. Pre-heat the oven to 220°C/425°F (gas mark 7). Brush the dough with freshly squeezed orange juice. Bake for 40 minutes, then remove from the tin and bake for a further 5 to 10 minutes. Allow to cool completely on a wire rack before slicing.

Kirribilli raisin plait

25g (1oz) dates
30cl (½ pint) skimmed milk
2.5cm (1 inch) piece vanilla
    pod (or 1 tsp vanilla
    essence)
1 egg
6 tbsp sunflower oil
1 tsp dried yeast
250g (8oz) wholewheat flour

175g (6oz) strong white flour
25g (1oz) wheat germ
25g (1oz) soya flour
150g (5oz) raisins
Grated rind of ½ orange and
    ½ lemon
1 tsp mixed spice
½ tsp cardomom powder
½ tsp grated nutmeg

Soften the dates by warming gently with the milk and vanilla pod (or essence). Remove vanilla pod and liquidize milk and dates with the egg and sunflower oil. Disperse the yeast in the mixture and set aside.

Mix all the remaining ingredients. Add the yeast mixture and knead to a soft dough. Cover with a damp cloth and leave in a warm place to rise. Re-knead after an hour or so. Divide into three equal parts and hand roll into three long sausages. Join by pressing together gently at one end and plait loosely. Seal the ends by pressing together. Brush with skimmed milk, cover and leave to rise in a warm place for a further hour.

Bake at 200°C/400°F (gas mark 6) for 30 minutes. Cool on a wire rack.

Best eaten fresh.

Carrot bread

1 tsp dried yeast
15cl (¼ pint) warm water
6 tbsp unsweetened orange
    juice
4 tbsp sunflower oil
1 egg (optional)
250g (8oz) wholewheat flour

250g (8oz) strong white flour
250g (8oz) grated carrot
1 tbsp cinnamon
small pinch salt
2 tsps orange, lemon &
    grapefruit rind, grated

Disperse the yeast in warm water and orange juice for ten minutes. Add the oil and beaten egg if desired.

Mix all the remaining ingredients. Stir in the yeast mixture, blending thoroughly. Cover and leave to rise in a warm place for one hour. Tip the risen dough into an oiled, lined 1kg bread tin (approx. 2lb). Cover and leave to rise for a further hour.

Bake at 220°C/425°F (gas mark 7) for 40 minutes and for a further 10 minutes out of the tin at 180°C/350°F (gas mark 4).

Cool on a wire rack before cutting.

Banana bread

1 tsp dried yeast
2 tbsp warm water
2 very ripe bananas (about 250g/8oz)
4 tbsp sunflower oil
6 tbsp yoghurt
175g (6oz) strong white flour

250g (8oz) wholewheat flour
grated rind of ½ lemon and ½ orange
2 tsp cinnamon powder
50g (2oz) chopped walnuts
40g (1½oz) raisins or sultanas

Dissolve the yeast in warm water. Liquidize the bananas, oil and yoghurt. Add the yeast mixture and set aside. Blend the flours (reserving 125g/4oz of wholewheat flour for later), rinds and cinnamon. Stir in the banana and yeast mixture. Cover and leave in a warm place to rise for 1 hour. Fold in the raisins, walnuts and remaining wholewheat flour and tip the mixture into an oiled, lined 1kg (2lb) bread tin. Cover and allow to rise for a further 30 minutes. Bake at 220°C/425°F (gas mark 7) for 40 minutes.

The cooked bread is golden brown. Remove from the tin and allow to become quite cold before slicing.

Cream cheese tartlets
(makes 18)

Pastry
125g (4oz) strong white flour
40g (2½oz) wholewheat flour
15g (½oz) soya flour
15g (½oz) wheat germ
1 tsp baking powder
4 tbsp sunflower oil
6 tbsp cold water
1 tsp rose water (optional)

Filling
300g (10oz) skimmed milk
   soft cheese
1 egg
40g (1½oz) seedless raisins
grated rind of 1 orange

To make the pastry blend the flours, wheat germ and baking powder. Rub in the sunflower oil to breadcrumb consistency and bind with cold water and rose water if desired. Roll out thinly on a floured board and line 18 lightly oiled patty tins with the pastry.

Blend the soft cheese and egg until smooth. Wash, drain and chop the raisins. Add these and the rind to the cheese. Fill each pastry case with about 1 dessertspoonful of mixture. Bake for 20–25 minutes at 220°C/425°F (gas mark 7). Remove from the tins and serve hot or cold.

Hermits
(makes about 24)

150g (5oz) wholewheat flour
300g (10oz) strong white flour
2 tbsp wheat germ
2 tbsp soya flour
2 tsps baking powder
2 tsps cinnamon powder
2 tsps grated nutmeg
175g (6oz) chopped mixed
   nuts

250g (8oz) finely grated carrot
Grated rind 1 orange and 1
   lemon
150g (5oz) currants
12.5cl (4fl oz) sunflower oil
¼ litre (8fl oz) unsweetened
   orange juice

Mix all the ingredients together. Drop spoonfuls onto an oiled baking sheet some distance apart and bake at 180°C/350°F (gas mark 4) for 25–30 minutes. Cool on a wire rack and store in an airtight tin.

Fruit scones

175g (6oz) strong white flour
and 1 tsp baking powder
or
175g (6oz) 81% self raising
flour

50g (2oz) porridge oats
90g (3oz) sultanas
4 tbsp sunflower oil
skimmed milk

Mix all the dry ingredients except the sultanas. Rub in the oil. Add the sultanas and bind all together with a little milk to form a soft dough. Roll out gently on a floured board to about 2.5cm (1 inch) thick. Cut into rounds and bake on a greased baking sheet at 220°C/425°F (gas mark 7) for 20 minutes.

Savoury drop scones

4 tbsp strong white flour and
1 tsp baking powder
or
4 tbsp 81% self raising flour
skimmed milk

pinch salt, pepper and
cayenne pepper
1 dessertspoon chopped
parsley

Mix all the dry ingredients, then add enough milk to make a thick creamy batter. Heat oil in a frying pan. Drop a teaspoonful of batter at a time into the hot oil. Cook on both sides until golden. Serve immediately.

Add well-crisped chopped pieces of bacon to the batter as an alternative to parsley.

Oatmeal biscuits
(makes approx 60)

50g (2oz) sultanas
6 tbsp skimmed milk
6 tbsp sunflower oil

175g (6oz) wholewheat flour
1 tsp baking powder
150g (5oz) medium oatmeal

Warm the sultanas with the milk. Liquidize with the oil and set aside to cool. Blend the flour, baking powder and oatmeal. Stir in the sultana mixture to form a soft dough. Roll out thinly on a well floured board. Cut into shapes and bake on an oiled baking sheet at

190°C/375°F (gas mark 5) for 12–15 minutes. Cool on a wire rack and store in an airtight container.

Johnny Cake

Give me some Johnny Cake, thick, thick, thick,
And a piece of butter, quick, quick, quick,
   Peter stands at the gate
   With his knife and his plate
Waiting for butter to put on his Johnny Cake.

Johnny Cake is the sugarless precursor of the gingerbread man and is of great antiquity. Like the Mexican *tortilla* and the Indian *chappatti*, it has its roots in the neolithic flat breads baked on hot stones at the fire's edge. In the southern United States small cakes of it are used as a side dish for savoury courses such as spare ribs. Historically it may well have provided compact and useful hard tack for the traveller, hence the suggestion that 'jonnycake' is a corruption of journey cake, though it may also derive from the American Indian shawnee-cake.

Florence White quotes Lady Gomme of the English Folk Cookery Association explaining the legend which we now associate with the gingerbread man complete with currant buttons and eyes, edging free of the oven and the cartwheels of the never-to-be-caught escaping boy.[9]

Here is our adaptation.

150g (5oz) plain white flour	4 tbsp sunflower oil
150g (5oz) maize (or Indian corn) meal	1 egg
	6 tbsp skimmed milk
1 tsp baking powder	currants

Blend the flour, meal and baking powder. Rub or work in the oil. Then mix to a stiff paste with the egg and milk. Roll out on a floured board to about 1cm (½ inch) thick and cut into little men (or women!) Decorate with currant eyes and buttons and bake on an oiled baking sheet for 20–25 minutes at 200°C/400°F (gas mark 6). Eat while still warm, on its own, with a spread of peanut butter or with grated cheese.

An equally good but lighter version requires 15cl (¼ pint) extra milk to form a thick batter. Drop spoonfuls into oiled patty tins and bake as before, until golden.

A gingerbread man and forty little biscuits!

125g (4oz) dates
4 tbsp water
4 tbsp unsweetened orange
  juice
6 tbsp sunflower oil
125g (4oz) strong white flour
and 1 tsp baking powder

or

125g (4oz) 81% self-raising
  flour
150g (5oz) wholewheat flour
50g (2oz) oatmeal
2½ tsp ground ginger

Soften the dates by warming in a pan with the water. Liquidize with the orange juice and oil. Blend the remaining ingredients, stirring in the date mixture to form a fairly soft, slightly sticky dough. Roll out thinly on a well floured board. Cut a gingerbread man. The remaining dough will make about 40 biscuits. Bake on an oiled baking sheet for 15–20 minutes at 200°C/400°F (gas mark 6). Cool on a wire rack and store in an airtight container.

Fruit and almond biscuits
(makes approx. 50)

75g (2½oz) raisins
5 tbsp skimmed milk
4 tbsp sunflower oil
125g (4oz) wholewheat flour

1 tsp baking powder
50g (2oz) medium oatmeal
50g (2oz) rice flour
40g (1½oz) ground almonds

Soften the raisins by warming with the milk. Liquidize with the oil and set aside to cool. Blend the remaining ingredients and stir in the raisin mixture to form a soft dough. Roll out thinly on a floured board. Cut into shapes and bake on an oiled baking sheet for 15–20 minutes at 200°C/400°F (gas mark 6). Cool on a wire rack and store in an airtight container.

Carrot Crackers
(makes approx 60)

250g (8oz) wholewheat flour
125g (4oz) strong white flour
1½ tsp baking powder
6 tbsp sunflower oil
175g (6oz) grated carrot

½ tsp salt (optional)
1½ tsps cinnamon powder
¼ tsp cumin powder
¼ tsp ground cloves
4 tbsp cold water

Mix the flours and baking powder and rub in the oil to breadcrumb consistency. Add grated carrot, salt (if desired) and spices. Mix to a fairly firm dough with the water. Roll out thinly on a floured board. Cut into shapes and bake on an oiled baking sheet at 190°C/375°F (gas mark 5) for 20–25 minutes. Cool on a wire rack and store in an airtight container.

Spicy orange biscuits
(makes approx. 50)

125g (4oz) dates
Juice of ½ lemon
6 tbsp orange juice
  (unsweetened)
6 tbsp sunflower oil
150g (5oz) wholewheat flour

150g (5oz) strong white flour
1 tsp baking powder
1 tbsp soya flour
25g (1oz) medium oatmeal
2½ tsps ground ginger
Grated rind of 1 orange

Soften the dates by warming with the lemon and orange juices. Liquidize with the oil and set aside to cool. Blend all the remaining ingredients, stirring in the date mixture to form a soft, sticky dough. Roll out thinly on a well floured board. Cut into shapes and bake on an oiled baking sheet at 220°C/425°F (gas mark 7) for 12 minutes. Cool on a wire rack and store in an airtight tin.

Cheese puffs
(makes approx. 36)

125g (4oz) wholewheat flour
1 tsp baking powder
good pinch cayenne pepper

175g (6oz) finely grated Edam
  cheese
4 tbsp water

Season the flour and baking powder with cayenne pepper and blend in the grated cheese. Bind with cold water to make a soft dough. Roll out thinly on a floured board and cut into rounds. Bake on an oiled baking sheet at 220°C/425°F (gas mark 7) for 12–15 minutes. Cool on a wire rack and store in an airtight container.

Cheese sticks
(makes approx. 50)

175g (6oz) wholewheat flour
1 tsp baking powder
Cayenne or freshly ground
   black pepper

125g (4oz) grated Edam
   cheese
4 tbsp sunflower oil
2 tsps mustard
1 egg white

Season the flour and baking powder with pepper and blend in the cheese. Whisk the oil, mustard and egg white together and stir into the other ingredients until thoroughly blended. Roll out on a floured board to about ½cm (¼ inch) thick. Cut into thin sticks and bake on an oiled baking sheet for 15 minutes at 200°C/400°F (gas mark 6). When cold, store in an airtight container.

Cocoa cashew spread

25g (1oz) dates
4 tbsp water
125g (4oz) cashew nuts

2 tsp cocoa powder
1 tbsp oil

Soften dates by warming gently with the water. Grind the nuts finely. Add cocoa powder, dates and oil. Blend and store in screw top jars in the refrigerator.

Banana spread

1 ripe banana
juice of ½ lemon

1 tbsp peanut butter
(see page 93)

Mash the ingredients together. This spread will not keep and should be used immediately.

Fig and cream cheese whip

6 dried figs
90g (3oz) skimmed milk
   soft cheese

4 tbsp orange juice
1 fresh orange (peeled, pipped
   and diced)

Cover the figs with boiling water and allow to soften overnight. Drain and chop coarsely. Gradually add orange juice to the cheese and whip until smooth. Fold in diced orange and chopped figs. Chill and use as required. This spread will keep for several days in the refrigerator.

Cheese spread

125g (4oz) grated hard mature
   cheese
1 tbsp skimmed milk powder

6 tbsp water
6 tbsp cooking oil
1 tsp anchovy essence

Dissolve the dried milk in water in a saucepan. Add the oil and anchovy essence and warm gently. Liquidise with the cheese until thick and creamy. Store in an airtight container in the refrigerator. Remove from the refrigerator 30 minutes before using. Keeps for about three weeks.

The advantage of this recipe is that it is considerably cheaper than the commercial product.

10 Just desserts
Puddings and sweets

Such eating, which the French call desert,
is unnaturall.

W. Vaughan

One solid dish his weekday need affords,
An added pudding solemnized the Lords.

Pope

Whatever we call them, puddings, sweets or desserts, they all owe their dominance to the ability of cheap sugar to create a visual fantasy world wherein the appeal of food is not taste or nutritive value, but its physical appearance. The mother who gurgles at her baby 'I could gobble you up', and the sexual partner who 'looks good enough to eat' remind us that this link between palatability and physical appearance is fairly fundamental. Hence the pull of the food pornographer's art which leers at us from every bookstall.

Against the fantasy the realities of actual taste often let us down with a bump. Who, after all, actually enjoys eating those depressing little pieces of wedding cake that circulate after the event? The ritual nibble that follows the dismantling of the sugary skyscraper is rarely pleasurable. The essence of the pastrycook's art is to persuade us to eat, not only when we are not hungry, but when our stomachs are already exhausted by a good meal. It is the glaze of the open tart, the featheriness of the sponge, the lightness of the meringue which tempt the glutton in us. These qualities come from sugar's ability to trap air, seal in flavour and stretch protein, and the sugar has to come in quantity. It also comes both as calories and as plaque food and, moreover, the sheer amount of the stuff in conventional recipes smothers all subtlety with its paralyzing blaze of undifferentiated sweetness.

That being established there have to be qualifications. We do have tastebuds designed to apprehend sweetness and they don't get much of a look in from the earlier courses of a meal. It is therefore refreshing to complete a meal with a fruit or a salad of mixed fruits. For adults nothing more should be necessary. For ravenous and active children a different strategy is required. The pud, sweet or dessert is well lodged into our culture and it makes good sense to work on the details rather than attempt to dislodge it. What we have done, as ever, is to fillet out the bag sugar which has come to dominate the pudding recipe and start rebuilding on more sensible lines. None of the recipes in this chapter makes use of sugars, syrups, honeys or any other concentrate. We occasionally permit a somewhat higher natural sugar to cereal ratio than that featured in the preceding chapter but we are still well below the orthodox levels. In none do we exceed the chosen limit of fresh fruit sweetness, even when this is achieved through the use of dried fruits.

The possibilities are endless and the following list should function, not merely as a repertoire in itself, but as a range of 'pointers' to illustrate new possibilities to the adventurous cook. It includes not only 'real' puddings for the hungry child but the more delicate sort for dinner parties. In both cases the scientifically sensible habit of completing the meal with cheese should be cultivated. In this way we escape the conventional wisdom's requirement to clean teeth after meals though we all know deep down that the death rate in the stampede from rice pudding to washbasin was always exaggerated. For children it might take the form of cheese shapes, small chunks of mousetrap, or the more sophisticated delights of the cheeseboard. For the adults it provides a good excuse for opening another bottle of wine and this, in moderation, should take the place of sugary liqueurs and sweetened coffee, both of which are sure toothkillers. Unsweetened coffee or tea are, of course, dentally harmless.

THE SAUCES

Blackcurrant sauce 1

40g (1½oz) currants   1 tbsp lemon juice
6 tbsp water
125g (4oz) fresh or frozen
    blackcurrants
    (unsweetened)

Wash and drain the currants then soak overnight in the water. The next day wash and drain the blackcurrants. Liquidize with the currants, water and lemon juice. Strain and store in the fridge until required.

Blackcurrant sauce 2

3 tsp cornflour   250g (8oz) blackcurrants
30cl (½ pint) apple juice
    (unsweetened)

Mix the cornflour with a little apple juice in a thick-bottomed saucepan. Liquidize the blackcurrants with the remaining apple juice. Strain on to the cornflour mixture and bring slowly to the boil, stirring constantly. Cook until the sauce begins to clear. Serve hot or cold.

Whipped banana cream

1 ripe banana   1 tsp lemon juice
50g (2oz) ground almonds

Liquidize all the ingredients until smooth and creamy. This is an ideal topping for puddings. It should be made and used fresh as it tends to discolour slightly after a few hours.

Chocolate sauce

15g (½oz) dates
1 dessertspoon cornflour

1 tsp cocoa powder
30cl (½ pint) skimmed milk

Soften the dates by heating gently in a small saucepan with a little water. Blend together in the liquidizer the cornflour, cocoa, milk and softened dates. Strain the mixture into a saucepan. Just before serving the pudding, bring the sauce to the boil and cook for 3 minutes, stirring constantly.

Date sauce

25g (1oz) dates
30cl (½ pint) skimmed milk

1 tbsp cornflour

Soften the dates by heating with just enough water to cover. Liquidize all the ingredients. Strain into a saucepan. Just before serving the pudding bring the sauce to the boil and cook for 3 minutes, stirring constantly.

Banana custard

30cl (½ pint) skimmed milk
1 tbsp custard powder

1 ripe banana

Mix the custard powder to a smooth paste with a little of the cold milk. In a saucepan bring the remainder of the milk to boiling point. Pour over the custard paste, stir thoroughly and return to the pan. Bring to the boil stirring constantly. Liquidize the custard with the banana until smooth and creamy. Serve while hot.

Orange Cream

6 tbsp skimmed milk powder

12.5cl (4fl oz) unsweetened
orange juice

Dissolve the skimmed milk powder in the orange juice and chill. Use instead of cream.

THE PASTRIES

> . . . There be some that to this paste add sugar, but it is certain
> to hinder the rising thereof

Recipe circa 1600

A good pastry really can 'lift' any pie or tart. The rough puff pastry
is crisp and light. The short pastry is not! It is nevertheless very
tasty, low in fat and to be preferred as a general rule. It can be
varied by adding rose water, sesame seeds, orange peel and grated
carrot to complement the filling. Cooking times vary depending on
the fillings but usually 25 minutes at 220°C/425°F (gas mark 7) is
sufficient.

Rough puff pastry

125g (4oz) plain flour
50g (2oz) wholewheat flour
150g (5oz) butter or
   margarine

pinch salt
ice cold water

Blend the flours and salt in a mixing bowl. The butter should be
used straight from the fridge. Add to the flour, cutting into pieces
about 1cm (½ inch) square. Use enough cold water to bind the
pastry. On a well-floured board roll out a long strip about ½cm
(¼ inch) thick. Fold over and over from one end and 'rest' in the
fridge for 10 or 15 minutes. Roll out once more in the same way,
fold and return to the fridge for a further 10 minutes after which
the pastry will be ready for use. You will need plenty of flour for
rolling out, especially if you use margarine.

Low fat short pastry

125g (4oz) strong white flour
75g (2½oz) wholewheat flour
15g (½oz) soya flour
15g (½oz) wheat germ
1 tsp baking powder

4 tbsp sunflower oil
6 tbsp cold water
1 tsp rose water (optional)
Sesame seeds (optional)

Blend the flours, wheat germ and baking powder. Rub in the sunflower oil to breadcrumb consistency and bind with cold water and a teaspoonful of rose water if desired. Sprinkle baking tins with sesame seeds before laying in the pastry. For savoury tarts and pies omit the rose water.

Apple tart

250g (8oz) rough puff pastry
2½ tsp cornflour
30cl (½ pint) apple juice

250g (8oz) cooking apples
(peeled, cored and thinly
sliced)
a little lemon juice

Line a 20cm (8 inch) flan case with pastry rolled out to about ½cm (¼ inch) thick. Chill in the fridge. Mix the cornflour with a little apple juice in a thick-bottomed pan. Add the remaining apple juice. Bring to the boil, stirring constantly and cook gently until the mixture thickens and clears. Remove from the heat and set aside to cool.

Peel and core the apples and drop into cold water into which some fresh lemon juice has been squeezed. This prevents browning and adds a tangy flavour. Next slice the apples as thinly as possible and arrange neatly in the pastry case. Pour the sauce evenly over the apples, making sure they are all coated.

Roll out any leftover pastry to about 2mm (⅛ inch). Cut into thin strips and lay across the tart criss-cross fashion, gently pressing down the edges. Bake on a baking sheet for 25 minutes at 220°C/425°F (gas mark 7). Reduce heat to 180°C/350°F (gas mark 4) for a further 10 minutes. Cool slightly on a wire rack before serving with fresh cream, natural yoghurt, or skimmed milk soft cheese.

Apple tarts with hot blackcurrant sauce

250g (8oz) pastry (see page 119)
4 medium sized cooking apples

blackcurrant sauce 2 (see page 117)

Roll out the pastry fairly thinly. Cut four rounds 10cm (4 inches) in diameter. Place these on a lightly greased baking sheet. Peel, core and thinly slice each apple. Arrange the slices neatly on each round of pastry. Bake for 25 minutes at 220°C/425°F (gas mark 7). Serve with hot blackcurrant sauce 2.

Apple pie baked with rose petals

> No flippant sugared notion
> Shall my appetite appease,
> Or bate my soul's devotion
> To apple-pie and cheese!
> *Eugene Field*

250g (8oz) pastry (see page 119)	rose petals
	40g (1½oz) sultanas
500g (1lb) cooking apples	2 whole cloves
a little lemon juice	zest of ½ lemon and ½ orange

Prepare the pastry and keep cool in the fridge.

Peel, core and cut each apple into eight pieces. Leave in a panful of cold water into which has been squeezed a little lemon juice. Simmer peel and cores in 30cl (½ pint) of water for 20 minutes. Choose a fragrant rose in full bloom. Remove the petals and set aside. Wash and drain the sultanas. Strain the apple stock and allow to cool.

Pile the apples into a pie dish, sprinkle with sultanas and put in two whole cloves, one at each end of the pie. Grate the zest of orange and lemon over the apples and cover with a good layer of rose petals. Half-fill the pie dish with cooled apple stock. Roll out the pastry to 1cm (¾ inch) thick. Cut a thin strip for the edge of the dish. Wet the edge of the dish with apple stock to make sure the pastry sticks. Wet the upper side of the strip with stock and lay the pastry crust on top. Crimp and trim the edge with a knife. Bake at 220°C/425°F (gas mark 7) for 25 minutes. Reduce to 170°C/325°F (gas mark 3) for a further 20 minutes. Serve hot with fresh cream, yoghurt, or skimmed milk soft cheese.

Apple and carrot pie

500g (1lb) pastry (see page 119)
300g (10oz) cooking apples
(peeled, cored and sliced)
75g (2½oz) finely grated
carrot
40g (1½oz) sultanas

20g (¾oz) dried banana,
finely sliced
grated rind of ½ orange and
½ lemon
cinnamon powder
fresh unsweetened orange
juice or skimmed milk

Line a lightly oiled pie dish, about 20cm (8 inches) in diameter
with rather more than half the pastry. Fill the case with layers of
apple and carrot sprinkled with the sultanas, dried banana, grated
rinds and cinnamon at intervals. Cover with the remaining pastry.
Seal the edges, brush the top with orange juice or skimmed milk.
Make a slit in the centre and bake at 200°C/400°F (gas mark 6) for
40 minutes.

PROPER PUDS

Baked apple and raisin pudding

17.5cl (6fl oz) unsweetened
orange juice
1 tsp dried yeast
1 eating apple
50g (2oz) wholewheat flour
50g (2oz) rice flour
15g (½oz) soya flour

50g (2oz) porridge oats
50g (2oz) ground almonds
1 tsp mixed spice
½ tsp cinnamon powder
3 tsp oil
125g (4oz) raisins (or sultanas)
zest of ½ orange and ½ lemon

Warm the orange juice to blood heat. Dissolve dried yeast in
liquidizer goblet with warmed orange juice. Wash and roughly
chop the whole apple. Add to the yeast and orange juice and
liquidize until the apple is reduced to pulp. Set aside.

Mix all remaining ingredients in a bowl. Add the yeast mixture
and stir to a creamy consistency. Pour into a greased pudding basin
and bake at 190°C/375°F (gas mark 5) for 45 minutes. Turn out on
to a dish and serve with hot chocolate sauce (see page 118).

Roly poly fruit pudding

90g (3oz) dried apricots
250g (8oz) cooking apples,
  peeled and cored
50g (2oz) sultanas
breadcrumbs or unsweetened
  cake crumbs (see Yoghurt
  cake, page 103)

25g (1oz) dried banana
25g(1oz) margarine or 2 tbsp
  sunflower oil
250g (8oz) 81% self-raising
  flour
15cl (¼ pint) skimmed milk

With just sufficient water to cover, simmer the apricots until tender (about 30 minutes). Chop the apples and add these and the sultanas to the apricots. Bring to the boil, remove from the heat and cool.

Rub the margarine or oil into the flour. Mix to a soft dough with the milk. Roll out on a floured board to make a rectangle 45×22.5cm (18×9 inches) and ½cm (¼ inch) thick. Sprinkle with bread or cake crumbs and spread with the fruit mixture. Dot with thinly sliced dried banana. Starting at one end, roll the dough and fruit loosely, rather like a Swiss roll. Bake in a greased rectangular pie dish or bread tin at 190°C/375°F (gas mark 5) for 30 to 40 minutes. Serve hot with date sauce (see page 00).

Banana sausages or bananes en croûte
(serves 6)

3 ripe bananas
juice of ½ orange and ½
  lemon
500g (1lb) pastry (see page
  119)

ground cinnamon to taste
grated zest of orange and
  lemon
a little skimmed milk

Cut the bananas in half lengthways. Soak in the juice and set aside. Roll out a rectangle of pastry about 2mm (⅛ inch) thick. Cut into six and place a piece of banana on each. Sprinkle with cinnamon and a little orange and lemon zest. Fold and stretch the pastry over the banana, sealing the edges with a little milk. Brush the tops with milk and bake on a greased baking sheet for 30 minutes at 220°C/425°F (gas mark 7). Serve hot with fresh cream, yoghurt or skimmed milk soft cheese.

Apple and almond pudding

50g (2oz) ground almonds
25g (1oz) Shredded Wheat
    crumbs
500g (1lb) cooking apples,
    peeled, cored and thinly
    sliced
grated zest of ½ orange and
    ½ lemon

40g (1½oz) dried apricots,
    pre-soaked
40g (1½oz) sultanas, washed
30cl (½ pint) skimmed milk
1 egg
1 ripe banana

Lightly oil an ovenproof dish. Mix together the ground almonds and crumbs. In the dish arrange layers of crumbs, apples, a sprinkling of zest and dried fruits until all are used up. Finish with a layer of crumbs.

To make the custard, bring the milk to the boil. Cool and liquidize with the egg and banana. Pour over the pudding mix, cover with greaseproof paper to prevent burning and bake at 190°C/375°F (gas mark 5) for 40 minutes. The top should be crisp and golden. Serve with fresh cream or yoghurt.

Blackberry and apple crumble

500g (1lb) cooking apples
    (peeled and cored)
a little lemon juice
Crumble topping
90g (3oz) mixed oats, barley
    and wheat flakes

15g (½oz) desiccated coconut
15g (½oz) ground almonds
250g (8oz) blackberries
1 tbsp sunflower seeds
    (optional)
4 tbsp cooking oil

Chop the apples and put into a pan of cold water into which a little lemon juice has been squeezed. Simmer the peel and cores in 30cl (½ pint) of water for 20 minutes. Strain the stock and cool.

To prepare the crumble topping, mix all the dry ingredients. Add the oil and mix until all ingredients look coated.

Drain the apples and pile into an ovenproof dish with the blackberries. Half fill the dish with stock and cover with crumble topping. Bake for 20 minutes at 220°C/425°F (gas mark 7) then reduce to 180°C/350°F (gas mark 4) for a further 20 minutes. Serve with fresh cream, yoghurt or skimmed milk soft cheese.

Apricot and apple crumble

50g (2oz) dried apricots
crumble topping (see
blackberry and apple
crumble above)

500g (1lb) cooking apples,
peeled and cored
25g (1oz) sultanas, washed
15g (½oz) diced dried
bananas

With just enough water to cover, simmer the apricots until tender (about 30 minutes). Prepare the crumble and set aside. Chop the apples, add to the apricots and bring to the boil. Tip into an ovenproof dish and mix in the sultanas and dried banana. Cover with crumble and bake at 220°C/425°F (gas mark 7) for 20 minutes. Reduce to 180°C/350°F (gas mark 4) and bake for a further 20 minutes.

Bread and butter pudding

6 thin slices of bread
25g (1oz) sultanas, washed
15g (½oz) diced dried banana
zest of ½ orange and ½ lemon
45cl (¾ pint) skimmed milk

1 beaten egg
bay leaf
2.5cm (1 inch) piece vanilla
pod

Lightly oil an ovenproof dish. Cut each slice of bread into four and arrange in the dish in layers, sprinkled with dried fruit and zest. Finish with a layer of bread. Bring the milk to the boil and cool for 5 minutes. Pour on to the beaten egg and whisk. Tip the custard carefully down the inside of the dish. Push the vanilla pod into the middle and top with a bay leaf. Allow to stand for 1 hour. Bake at 180°C/350°F (gas mark 4) for 40 minutes or until the top is golden and the custard set.

Steamed Sponge Pudding

40g (1½oz) dates
3 tbsp water
1 egg
4 tbsp sunflower oil
12.5cl (4fl oz) skimmed milk
125g (4oz) wholewheat flour

90g (3oz) strong white flour
1 tsp baking powder
50g (2oz) currants and
   sultanas
Grated rind of ½ orange and
   ½ lemon

Soften the dates by warming gently in water and liquidize with the egg, oil and milk. Blend the remaining ingredients then add the date mixture, stirring vigorously. Steam in an oiled pudding basin, covered with greaseproof paper and tied with a cloth, for 1½ hours. Serve with date sauce (see page 118).

Spicy Steamed Pudding

50g (2oz) wholewheat flour
175g (6oz) strong white flour
2 tsp baking powder
15g (½oz) wheat germ
15g (½oz) soya flour
2 tsp mixed spice
1tsp ground ginger
250g (8oz) coarsely shredded
   or chopped apple (cored)

50g (2oz) finely grated carrot
25g (1oz) raisins
50g (2oz) chopped dried
   apricots
grated rind ½ orange and ½
   lemon
3 tbsp sunflower oil
12.5cl (4fl oz) low fat natural
   yoghurt

Thoroughly blend all the ingredients. Steam in a lightly oiled pudding basin covered with greaseproof paper and tied with a cloth for about 2 hours. Turn out onto a hot serving dish and serve with date sauce (see page 118).

Baked apples

4 large cooking apples
4 dates
½ sliced dried banana
sultanas and/or raisins

sunflower oil
cinnamon
water

Choose four good sized unblemished cooking apples. Remove the cores and arrange apples in a fireproof dish. Fill the middle of each with a date, two slices of dried banana and three or four sultanas. Brush the top with oil and sprinkle with cinnamon. Cover the bottom of the dish with water and bake at 190°C/375°F (gas mark 5) for 30 to 40 minutes.

Baked apples stuffed with fruits

4 dried figs	2 pears
4 large unblemished cooking apples	zest of ½ orange and ½ lemon
	apple juice
lemon juice	sunflower oil

Wash the figs and soak, either overnight in cold water or for about 30 minutes in boiling water, until they are plump and tender. Wash the apples and slice the tops off each. Scoop out the pulp to form cups with a wall about ¼ inch thick. (A grapefruit knife is useful.) Reserve the pulp in water to which a squeeze of lemon juice has been added. Drain the figs, chop roughly and divide between the four hollowed out apples. Peel core and chop the pears and divide these likewise. Drain enough pulp to fill up each cup adding a little zest of orange and lemon and enough apple juice to half fill the cups. Top with two or three fresh rose petals and replace the lids. Brush with oil and bake in an ovenproof dish for 40 minutes at 220°C/425°F (gas mark 7). Lift each lid and garnish with a blob of thick cream or yoghurt before serving.

Rice pudding

175g (6oz) brown short grain rice	15g (½oz) margarine (optional)
25g (1oz) sultanas (or raisins)	grated nutmeg
	90cl (1½ pints) skimmed milk

In a pudding basin wash the rice and sultanas with boiling water. Drain. Put a knob of margarine, if used, in the middle and sprinkle with nutmeg. Bring the milk to the boil and add to the rice. Bake at 150°C/300°F (gas mark 2) for 3 hours.

Baked semolina pudding

60cl (1 pint) skimmed milk
4 tbsp wholewheat semolina
25g (1oz) sultanas

1 egg
grated nutmeg

Warm the milk in a saucepan and stir in the semolina. Continue stirring and bring to the boil. Cook for 3 minutes, add the washed sultanas and cool for 15 minutes. Beat the egg and stir into the semolina. Pour into an oiled pudding basin, sprinkle with nutmeg and bake for 45 minutes at 170°C/325°F (gas mark 3).

Blackberry jelly

250g (8oz) blackberries
45cl (¾ pint) apple juice
  (unsweetened)

15g (½oz) gelatine

Cook the blackberries in the apple juice for ten minutes. Liquidize the fruit and juice. Dissolve the gelatine in a little hot water and add to the fruit purée. Liquidize again and strain into a mould. Cool and refrigerate until set. Serve with fresh cream or natural yoghurt.

Orange jelly

60cl (1 pint) unsweetened
  orange juice

25g (1oz) gelatine
1 tsp lemon juice

Heat the orange juice to boiling point. Dissolve the gelatine in a little hot orange and add to the rest. Stir in the lemon juice and pour into a mould or glass dish. Cool and refrigerate until set. Decorate with segments of orange, fresh mint leaves and cream.

Raspberry and apple kissyel (blancmange)

500g (1lb) cooking apples,
  cored
¾ litre (1¼ pints)
  unsweetened apple juice

500g (1lb) raspberries
5 level tbsp potatoflour

Wash and quarter the apples. Bring to the boil with 60cl (1 pint) apple juice. Cook until almost tender. Add the raspberries and bring to the boil once again. Liquidize the fruit and stock and strain into a thick-bottomed pan. Blend the cornflour with the remaining apple juice and add to the fruit purée. Bring to the boil, stirring constantly. Cook for ten minutes or until the opaqueness disappears. Pour into a wet mould, cool and chill. Turn out on to a dish and serve with freshly sliced apples, oranges and bananas.

ICES

Orange ice

4 heaped tbsp skimmed milk powder	60cl (1 pint) unsweetened orange juice
	4 tsp glycerin

Dissolve the milk powder in the orange juice. Add the glycerin, whisk and pour into a plastic container, seal and freeze for 30 minutes. Whisk thoroughly and freeze for a further 30 minutes. Whisk again and freeze for a final 30 minutes before serving.

If the ice is not to be eaten immediately, transfer it to the fridge 30 minutes before serving. The slight rise in temperature will improve its texture and flavour.

Much of the labour of making ices can be avoided by the use of one of the many excellent ice-cream makers on the market.

Strawberry ice

45cl (¾ pint) unsweetened apple juice	250g (8oz) strawberries
2 tbsp cornflour	2 tbsp skimmed milk powder
	2 tsp glycerin

Thicken 30cl (½ pint) of apple juice with the cornflour. Cook for 2 or 3 minutes then liquidize with the remaining apple juice and strawberries. Dissolve the milk powder in the mixture. Blend in the glycerin and freeze (see above for method).

Blackberry ice

250g (8oz) blackberries
45cl (¾ pint) unsweetened
  apple juice

5 tbsp skimmed milk powder
3 tsp glycerin

Cook the blackberries in the apple juice for 5 minutes. Liquidize
and leave to cool. Add the skimmed milk powder and glycerin,
strain into a freezer container and freeze (see above for method).

Blackberry and apple ice

500g (1lb) cooking apples,
  peeled, cored and sliced
250g (8oz) blackberries

¼ litre (8fl oz) unsweetened
  apple juice
5 tbsp skimmed milk powder
4 tsp glycerin

Cook the fruit in the apple juice until tender. Liquidize and leave
to cool for 15 minutes. Add the skimmed milk powder and
glycerin and strain into a freezer container. Freeze (see above for
method).

Remove from the freezer 30 minutes before serving.

THE DAINTY DISHES

The dividing line between a pudding and a dinner party delicacy is
not a clear one. Although baked semolina doesn't seem to go with
the Chateau d'Yquem, there is no reason to deprive children of the
experience of good food. The following, therefore, while appropriate for all, have sufficient style to garnish a more special occasion.

Although gelatine seems the most common setting agent there
are other methods which may suit vegetarians and those who find
its taste unpleasant. These are the vegetable gelatines extracted
from seaweeds, such as carrageen, otherwise known as Irish moss,
and agar-agar (Japanese or Ceylon moss). Agar-agar is sold under
the Sunwheel label and is available loose from wholefood shops.

Chocolate prune mousse

125g (4oz) prunes
250g (8oz) cooking apples
(cored)
15g (½oz) creamed coconut
2 tbsp skimmed milk powder
1½ tbsp cocoa powder

15g (½oz) gelatine
1 egg white
Whipped banana cream (page 117)
Walnuts or almonds for decoration

Wash and soak the prunes overnight in 30cl (½ pint) cold water. Discard the stones and cook the prunes gently with the apples. Meantime, dissolve the creamed coconut in 4 tablespoons of hot water, then stir in the skimmed milk powder as the liquid begins to cool. Liquidize the fruit and its juice. Add the coconut mixture and cocoa powder and blend once again. Dissolve the gelatine in 3 tablespoons of hot water. Stir into the fruit mixture and allow to become quite cold, but not set, before folding in the stiffly beaten egg white. Refrigerate until set and decorate with whipped banana cream and almonds or walnuts.

Apricot and almond pudding

125g (4oz) dried apricots
25g (1oz) raisins
250g (8oz) cooking apples
(cored)
4 tbsp unsweetened orange juice

1 tbsp lemon juice (optional)
15g (½oz) gelatine
50–90g (2–3oz) ground almonds
1 or 2 egg whites

Wash and drain the dried fruits. Simmer in 30cl (½ pint) water for 15 minutes. Add chopped apples and continue cooking until the latter are tender. Liquidize with the stock and juices. Dissolve the gelatine in 3 tablespoons of hot water and stir into the fruit purée. Allow to cool completely before folding in the ground almonds and stiffly beaten egg white(s). Refrigerate until set.

Topping variations: Reserve a little purée before adding almonds and egg white. Spread evenly over the finished pudding and scatter with flaked toasted almonds, grape halves and mint leaves.

Little chocolate soufflés

40g (1½oz) dates
30cl (½ pint) skimmed milk
2 tbsp skimmed milk powder
2 tbsp cocoa powder

15g (½oz) gelatine
4 tbsp hot black coffee
1 egg white

Soften the dates by warming gently with the milk. Liquidize and cool slightly before adding the skimmed milk powder and cocoa powder. Dissolve the gelatine in the hot coffee and add to the mixture. Cool completely before folding in the stiffly beaten egg white. Divide into four little dishes and refrigerate until set.

Coffee custard creams

40g (1½oz) dates
45cl (¾ pint) black coffee

5 tbsp skimmed milk powder
1 egg

Soften the dates in hot coffee, keeping it well below boiling point to preserve flavour. Cool slightly and stir in the skimmed milk powder. Cool for a further 10 minutes then liquidize with the egg. Strain and divide between four individual soufflé dishes. Place in a shallow vessel containing 2.5cm (1 inch) of water and bake for 1 hour at 150°C/300°F (gas mark 2). Refrigerate before serving.

Blackberry and apple soufflé

250g (8oz) blackberries
500g (1lb) cooking apples
  (cored and chopped)
75g (2½oz) sultanas
15cl (¼ pint) water
1 tbsp agar-agar or 15g (½oz)
  gelatine

5 tbsp skimmed milk powder
4 tbsp apple juice
15cl (5fl oz) low fat natural
  yoghurt (optional)
1 egg white
A few whole blackberries and
  mint leaves to decorate

In a tightly covered pan cook the blackberries, apples and sultanas with water until tender. Drain the juices into a saucepan. Liquidize the fruit and set aside. Bring the juice to the boil. Sprinkle on the flakes of agar-agar. Stir and cook for 5 minutes.

Add to the fruit purée and blend well. Dissolve the skimmed milk powder in apple juice and mix into the fruit purée. When quite cold but not set, stir in the yoghurt and stiffly beaten egg white. Refrigerate to set. Decorate with whole blackberries and mint leaves.

Gooseberry mould

300g (10oz) gooseberries
(topped and tailed)
500g (1lb) cooking apples
(cored and chopped)
75g (2½oz) sultanas
15cl (¼ pint) water

grated rind of ½ orange and
½ lemon
1 tbsp agar-agar or 15g (½oz)
gelatine
4 tbsp skimmed milk powder
4 tbsp unsweetened orange
juice

In a tightly covered pan cook the gooseberries, apples and sultanas with the water and grated rind until the fruit is tender. Drain the juice into a saucepan. Liquidize the pulp and set aside. Bring the juice to the boil. Sprinkle in the agar-agar. Stir and cook for 5 minutes. Blend with the fruit purée. Dissolve the skimmed milk powder in orange juice and blend with the fruit. Pour the mixture into a wet mould and when cool refrigerate to set. Turn out and decorate with slices of orange. Serve with orange cream (see page 118) or yoghurt.

Gooseberry cheese

To the above recipe add 175g (6oz) skimmed milk soft cheese when blending in the skimmed milk powder and orange juice. Set in a shallow dish and decorate with pieces of orange and mint leaves.

Raspberry trifle

175g (6oz) yoghurt cake (see
  page 103)
500g (1lb) fresh or frozen
  raspberries (unsweetened)
60cl (1 pint) unsweetened
  apple juice

2 tbsp sherry
15g (½oz) gelatine
30cl (½ pint) double cream or
  yoghurt

Cut the cake into 5cm (2 inch) squares or fingers and arrange in the
bottom of a glass bowl. Put 350g (12oz) raspberries on top with
30cl (½ pint) apple juice and the sherry. Set aside for 1 hour.
Strain the excess liquid into a jug. Heat to boiling point the
remaining apple juice. Dissolve the gelatine in hot juice and add to
the rest of the liquid. Cool for 15 minutes then pour over the fruit
and cake. When the jelly is cold, refrigerate the trifle until set.
Decorate with whipped cream or yoghurt and the remaining raspberries.

Sliced bananas and blackcurrant sauce

1 banana per person
thick cream or yoghurt

blackcurrant sauce (see p.117)

Slice one banana for each person. Pile into glass dishes and top
with blackcurrant sauce and a blob of cream, or yoghurt.

Grilled bananas

1 banana per person
cinnamon

sunflower oil

Slice the bananas lengthwise. Brush with oil and sprinkle with
cinnamon. Grill for 15 minutes and serve with cream or yoghurt.

11 What sauce!
Chutneys and ketchups

First note that in your bill of fare,
Sauce he provided for the rare
But vinegar the most extol;
'Tis of an hare the very soul.

Rabelais

. . . And so it came about that when the Ministry of Food
came to negotiate with the Ministry of War, the voice of
the generals carried considerable weight. And one thing
upon which the generals were adamant was that there
must be an adequate supply of gherkins. They were
unmoved to learn from the nutritional experts that
gherkins are – in terms of nutrients – of negligible
significance. The British soldier, they insisted, could not
fight without a proper supply of pickles to eat with his
cold meat.

Magnus Pyke
Food and Society

It may well be that the continuing British passion for condiment
sauces is the direct result of the British cooks' passion for baking to
tastelessness the meats they accompany. Whatever the reason, they
have a long tradition and, as we showed in Chapter 9, spices and
sauces made with them played an important role in the diet of our
ancestors. The functions they served were several. First was the
grim necessity to preserve food through the unproductive winter
months. Vinegar, salt, drying and smoking all played their part in
eking out reserves until the new summer could ensure fresh supplies. They also suppressed unfresh smells and flavours and from
such necessities proceeded culinary virtues. For even when food

was fresh the appetite for the sharp flavours remained undiminished. Now even with fresh ingredients available round the year, there is still a big market for the preserve.

The already considerable range of pickle and sauce available to the eighteenth-century household was given a fresh boost from overseas in the nineteenth. However aloof the imperial conquerors of India remained from the native populace, they fell victim to its culinary fragrances. The Indian magic of kedgeree, chutnee and curree thus added a further dimension to the tradition of combining and blending contrary flavours, sweet and sour, fruity and hot, sweet-sour and hot and so on. Sugar, though a natural presence through the spices, soon entered the scene in its own right. As spices and fruits remained dear and sugar became comparatively cheaper, it came to serve less as condiment and more as 'filler', while for the manufacturer concerned with shelf-life it became the chief ingredient. The end products of the historical development are then not so much condiment blends of contrasting flavours but a series of fruit and vegetable jams, set off with minute amounts of spice to supply a tiny signature of a lost authenticity.

It is not difficult, however, to restore the original balance, as the following recipes show. The first is a reconstruction of a sauce we bought in an old vodka bottle with a tarred cork at the free market in Batumi. Made from 'tkemali', a wild plum which grows freely in Trans-Caucasia, it accompanies the grilled meats and shashlyks traditional to that region.

Tkemali (plum) sauce

2kg (4lb) black stewing plums
15cl (¼ pint) water
15cl (¼ pint) carton natural
  yoghurt
1 tsp freshly milled pepper

4 cloves garlic (crushed or
  pressed)
¼ tsp cayenne pepper
3 sprigs fresh basil (tied in
  bunch) or 1 tsp dried basil

Stew the plums slowly in the water, stirring to prevent catching. When soft, cool and pass through a rice strainer. Add the rest of the ingredients and cook slowly on a low heat until the mixture

136

reaches ketchup consistency. Remove the fresh basil (if used) and pour hot into sterile jam jars, or, if it is to be frozen, cool and pour into margarine tubs. It should fill two 60cl (1 pint) jars or six margarine tubs.

Since the experiment we have discovered another and sweeter version from a native chef, Arto der Haroutunian, whose *Middle Eastern Cookery* was published by Pan in 1984. His tkemali uses prunes and it is worth notice that his other sauce recipes from the region are either sugar-free or sugar-low. Here it is.

Tkemali sauce
(traditional recipe)

450ml (¾ pint) water
225g (½lb) prunes
1 clove garlic
1 teaspoon ground coriander

½ teaspoon salt
½ teaspoon paprika
1½ tablespoons lemon juice

Bring the water to the boil in a saucepan, add the prunes, remove from the heat and set aside for 10 minutes. Bring back to the boil and cook briskly for about 15 minutes or until the prunes are tender. Pour the contents of the pan into a sieve placed over a bowl. Reserve the liquid.

Stone the prunes and put the flesh into a liquidizer with the garlic and coriander. Add a little of the reserved liquid and blend well. The sauce needs to have the consistency of thick cream and so adjust the amount of liquid you add accordingly. Transfer this sauce to a saucepan, stir in the salt and paprika and bring to the boil. Remove from the heat and stir in the lemon juice. This is usually served at room temperature.

Barbecue sauce

Deep down nobody really wants to plaster their lamb chops with sugar, though, if they use commercial barbecue sauce, that is what they will be doing. The home-made one is, as always, an improvement.

1 clove garlic (crushed)
1 large onion (finely chopped)
2 tbsp oil
4 tbsp tomato purée
2 tbsp lemon juice
salt, pepper to taste

15cl (¼ pint) meat or
  vegetable stock
4 tbsp mushroom ketchup
  ('Burgess' or home made)
2 level tsp ground mustard

Fry garlic and onion gently in oil until 'melted'. Stir in tomato purée, lemon juice, seasoning, stock, ketchup and mustard. Bring to boil and simmer for 15 minutes. Cool and store in refrigerator.

Few of us associate the word 'ketchup' with a brine made from Chinese pickled fish, though this is its origin. For most of us the word conjures up the red and slightly luminous slush in the supermarket, to which the manufacturers have added enough sugar to confer eternal shelf-life. The Chinese varieties, in a tradition common to other cultures, are created by the extraction and concentration of animal and plant ingredients. The Romans leaned very heavily in their recipes on *liquamen*. Our modern equivalent might be anchovy essence. It was prepared from small fish, sprats, anchovies and the like, mixed with the entrails of larger fish, and sun-dried. The liquor was strained off, perhaps in the Chinese

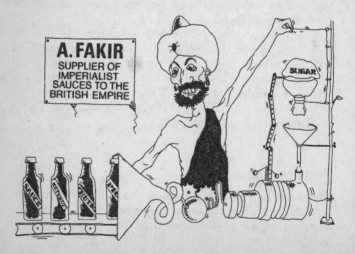

A. FAKIR
SUPPLIER OF
IMPERIALIST
SAUCES TO THE
BRITISH EMPIRE

manner, and stored in jars. It may well have been used as a substitute for salt, when the latter's supply could not be guaranteed.

These arts have not been altogether lost and sugarless walnut and mushroom ketchups are found in both recipe book and supermarket. Meanwhile we have to do something about commercial tomato ketchup which at 23% sugar must be one of our biggest turn-offs. We append two recipes with the comment that, while they may not have the astronomical life-span of the commercials, they are so good that the question never arises.

Tomato ketchup 1
(sufficient to fill two 500g (1lb) jam jars)

1.5kg (3lb) ripe tomatoes	1 clove garlic
1 medium onion	50g (2oz) dates
1 small sweet red pepper	15cl (¼ pint) wine vinegar

In a muslin bag

1 bay leaf	1 tsp mustard seeds
pinch chilli powder (optional)	5 peppercorns
1 tsp celery seeds or lovage seeds	piece of cinnamon (2 inches)

Chop the tomatoes (unpeeled), onion, pepper (including seeds), garlic and dates. Put all the vegetables into an enamel saucepan, cover with water and boil gently until soft (about 30 minutes). Liquidize, then strain back into the saucepan through a sieve. Meanwhile chop bay leaf and bruise the spices before tying into the muslin bag. Add bag and vinegar to the juice in the pan. Bring to the boil and simmer gently until the juice has reduced to ketchup consistency. (Test this by spooning a little juice on to a cold saucer from time to time.) Seal in hot sterilized jars. Once opened, refrigerate.

Tomato Ketchup 2

While the first recipe for ketchup is appropriate for the gardener who needs to preserve a surplus of tomatoes, most of us will turn to tins. The usual canned variety is the Italian plum-type and is sweet in its own right. The cheapest way to buy them is in large 850g (1¾lb) cans and the following recipe using three of them makes about 1.75kg (3½lb) of ketchup.

2 medium sized onions
1 large red pepper
2 cloves garlic

3×850g (1¾lb) cans tomatoes
15cl (¼ pint) wine vinegar

In a muslin bag
2 bay leaves
2 tsp lovage or celery seeds
2 tsp mustard seeds
10 peppercorns

5cm (2 inch) piece of
  cinnamon
¼ tsp chilli powder (optional)

Chop the onions, pepper (including seeds) and garlic. Put them into a large enamelled saucepan. Add the tomatoes and boil gently for about 1 hour or until the vegetables are soft. Meanwhile chop bay leaves, bruise lovage and mustard seeds, peppercorns and cinnamon, tie up the spice bag, and soak in the warmed vinegar for a few minutes. Liquidize the tomato mixture thoroughly and strain through a sieve back into the saucepan. Add the spice bag and vinegar to the juice and bring to the boil again. Simmer gently until the juice is reduced to ketchup consistency. (Test this from time to time by spooning a little on to a cold saucer.) Seal in hot sterilized jars. Vinegar bottles make good containers. Once opened, refrigerate.

Although the sugar cane arrived in India a good millenium before Europe became hooked, it seems, if one can judge from authentic chutney recipes, not to have had a dominant role on the sub-continent. The culinary transfer to Britain was accompanied by a takeover by the sugar culture. Just as we are reputed to like our sherry dry on the label but sweet in the bottle, our daring in embracing the powerful spices of the East was cushioned by sugar. Most commercial 'chutneys' therefore have sugar as the number one ingredient, even when their origins are allegedly Indian. Those which don't are listed in Appendix 1. As regards their home preparation, we shall not here attempt a study of those chutneys which traditionally accompany Indian food and readers must resort to the authentic sources. They will in the main be found to be sugarless. Here our concern is with those Anglo-Indian domestications which enabled the prudent Victorian housewife to use up surpluses of under-ripe fruits and to which sugar brought its baleful sameness. Given that we now have freezers and refrigerators, the importance of the sugar has diminished to zero and we make and eat them because we like them.

All of the following recipes survive the season without recourse to any other expedient than that of being on the pantry shelf. We start with that glorious product of wet old England, green tomato chutney.

Green tomato chutney
(to make 2.75kg (5½lb) chutney)

500g (1lb) cooking apples
1.5kg (3lb) green tomatoes
500g (1lb) onions
125g (4oz) sultanas (washed)
125g (4oz) chopped dates
2 cloves garlic (crushed)
45cl (¾ pint) wine or spirit
  vinegar

1 tsp chilli powder
3 tsp cumin powder
2 tsp ground ginger

Spice bouquet
1 bay leaf
½ tsp whole cloves
1 tsp crushed allspice

Peel and core the apples and wash the tomatoes. Chop tomatoes, apples and onions into small pieces and place in a large preserving pan. Add the dried fruits, crushed garlic and vinegar and bring to simmering point. Add the bouquet of spices, loosely cover and simmer for 2 hours when all the vegetables should be soft. Remove 15cl (¼ pint) of the mixture to a jug or mixing bowl and stir in chilli, cumin and ginger to form a thick paste. Stir this paste back into the pan and cook slowly for a further hour, stirring occasionally to prevent catching.

Spoon the hot chutney into warm sterile jars. Cover with a small disc of greaseproof paper and a further jam pot cover before sealing with the lid. Store in a cool place or refrigerate. Once open, keep in the refrigerator. This chutney freezes well and this is the best means of storage in the case of 'mass production.'

Plum and apple chutney

500g (1lb) black preserving
  plums or damsons
35cl (12fl oz) vinegar
125g (4oz) raisins (washed)
250g (8oz) cooking apples

175g (6oz) onions
½ tsp ground ginger
7g (¼oz) pickling spice (in a
  muslin bag)

Simmer the plums in vinegar until tender enough to remove the stones. This takes about 45 minutes, depending upon the nature and ripeness of the plums. When the stones have been removed, add the raisins and simmer for a further half hour. Add apple, onion, ginger and spice bag. Simmer until the onion and apple have softened and then until the mixture is thick. Spoon into hot sterile jars and seal. Store in a cool place. Once opened, refrigerate.

Pumpkin, tomato and ginger chutney

1kg (2lb) pumpkin (peeled
  weight)
250g (8oz) onions
750g (1½lb) ripe tomatoes
2 cloves garlic
50g (2oz) dates
2 tsps ground ginger

50g (2oz) raisins
45cl (¾ pint) vinegar (wine,
  cider, or spirit)

Spice Bouquet
2 tsp black peppercorns
2 tsp allspice berries

Dice the pumpkin into somewhat less than 1cm (½inch) squares. Chop onion, tomatoes, garlic and dates. Put all these into a preserving pan and add the ginger and raisins. Bruise the peppercorns and allspice in a mortar and tie into a gauze or muslin bag. Soak this in warmed vinegar for about 15 minutes before adding bouquet and vinegar to the pan. Place on a low heat until enough moisture has come out of the vegetables to prevent catching, then raise the heat until simmering starts. Maintain at simmering level for 1 hour. Spoon into hot sterile jars and store as for green tomato chutney.

Sweet banana chutney

500g (1lb) cooking apples
250g (8oz) onions
125g (4oz) sultanas
2 cloves garlic
½ tsp chilli powder
7g (¼oz) mustard seeds
    (bruised)

½ tsp turmeric
30cl (½ pint) vinegar (wine or
    spirit for preference)
500g (1lb) greenish bananas
    (peeled)

Chop the apples and onions finely, or pass through a coarse mincer. Put them in a thick saucepan with the sultanas, crushed garlic, spices and vinegar and bring slowly to the boil. Simmer gently for at least half an hour. It may be necessary to add small amounts of water to ensure that the onion has enough 'space' to cook and to prevent catching. Chop the bananas coarsely and add to the pan. Simmer and stir for a further half an hour. Remove from the heat and set aside for 30 minutes before putting into hot sterile jars and sealing. Store in a cool place. Once opened, refrigerate.

Hot banana chutney

To the ingredients above (sweet banana chutney) add one large green pepper, chopped or minced with the apple and onion, and a whole teaspoonful of chilli powder instead of half. This chutney is appropriate for curried dishes and it is difficult to prescribe fixed amounts of 'hot' ingredients. Those who like it hotter will add even more chilli and possibly garlic. The cooking and storing procedure is the same as above.

Baked beans

Finally, sugar-through-sauce is at its most significant in our diets through the ubiquitous 'baked beans in tomato sauce'. The beans are not of course baked but pressure-cooked in a sugary tomato liquid within the can. As a stand-by convenience food they have an important role and we can see no reason why a liking for beans should involve sugar at all. Average can content is about 6.1% sugar and we offer an alternative with none. It is not usually possible to obtain the hard American Navy beans which the canners use but ordinary haricot beans serve just as well and given the popularity of mushy peas, the slightly mushy beans of this recipe should go down well. Like all the bean recipes of southern Europe from which this is derived, it improves on reheating and freezes well.

Haricot beans in tomato sauce

500g (1lb) haricot beans
2 onions
1 medium carrot
½ red pepper
1 clove garlic
2 tbsp oil

1 × 400g (14oz) can tomatoes
   plus 1 tbsp tomato purée
or
500g (1lb) fresh ripe tomatoes
   plus 2 tbsp tomato purée
salt

Bouquet (tie in a muslin bag)

1 tsp lovage seeds
1 tsp white mustard seeds
½ tsp black peppercorns all
   crushed

½ tsp chilli powder
½ tsp basil
1 bay leaf

Soak the beans in cold water overnight. Drain and cover with fresh water. Bring to the boil and cook for 10 minutes. Discard the water. Simmer for 1 hour with 1.25 litres (2 pints) water, 1 onion, the carrot and the bouquet.

Finely chop the remaining onion, pepper and garlic and fry gently in oil until transparent. Add the tomatoes and tomato purée and cook for a further 15 minutes. Liquidize with the onion from the boiling beans and set aside. Remove the carrot and bouquet from the cooked beans. Add the tomato sauce and bring to the boil once more, simmering very gently for about 30 minutes. Season with salt to taste and freeze when quite cold.

12 Keeping the feast
Festal fare

We were young, we were merry, we were very, very wise
And the door stood open at our feast . . .

Mary Coleridge

Some eat the jelly baby whole but most
    Dismember it at leisure,
For, headless, there is no doubt that it gives
    A reasonable measure
    Of unexampled pleasure

Roy Fuller

That the body's sorcery can create its own living flesh from inert materials is essentially mysterious, and perhaps for this reason alone the world of food involves much more than mere nutrition. In matters of food choice and preference, rationality is a relatively inconspicuous contributor. Social custom, magic and mystery are interwoven with every passion, with every aversion. Honey, concentrated and disgorged from a myriad insect stomachs, has been accorded an exalted status in the food habits of Western man who views with horror the prospect of eating insects that other cultures, with good biochemical justification, regard as staple items of diet. Anyone born within the sound of Bow Bells will be no stranger to winkles and whelks but retch at the prospect of eating their landlubbing cousin, the snail.

Sometimes these dietary irrationalities acquire a veneer of scientific justification. The resemblance of a walnut's corrugation to the human brain once led to a view that intelligence could be increased by their consumption, while the colour of beetroot made it a certain cure for anaemia. This history of such naive correspondences is by no means ended. Newer sorceries have now given

145

magical status to such biological commonplaces as hormone, vitamin, enzyme, lecithin and so on.

The explosive advances of nutritional science made in the last decades have still left most people more ignorant of their bodies than their motor cars. It is hard to escape the view that much of this ignorance is an almost wilful rejection of the takeover bid that science is making for our innermost animal secrets. No matter how banal and unsensational the nutritionists try to make matters of diet, we seem to insist on a further dimension to clothe the ingestion of nutrients with atmosphere and spirit.

This is particularly obvious when we come to the more ritual and celebratory aspects of our eating. A spiced bun with a scaffold pressed into its surface may not differ in taste or nutritional value from its uncreased fellow, but the rationalisation will not convince Good Friday's child. The most nourishing Christmas pudding in the world still needs a darkened room, an atmosphere of tremulous excitement and a sprig of inedible holly to work its effect.

The climaxes giving structure to the primitive year derive from the basics of seedtime and harvest, plenty and want, solstice and equinox. When religion imposed its language and style on such events, the events themselves remained food-related. The winter solstice, for instance, coincided with a time when animal fodder was becoming scarce and the weaker beasts were slaughtered; allowing perhaps the last substantial meal until Spring. We may now call it Christmas and supposedly celebrate something else but we still eat as if there were nothing more to come in the months ahead.

Here lies a cosy nest for the sugar cuckoo, for if ever a lily was gilded, it is in that Christmas triad of pudding, cake and mincemeat. The recipes, densely sweet with dried fruits, still insist on a sugar quota, albeit that Trojan horse, the soft brown variety. So in our keeping and purging of the feasts we start with an easy one, where the added sugar was and is never necessary. There can be no doubt, of course, that with all the fruits' natural sugars and acid, the teeth will take a bit of a pasting. The plaque, however, will be getting its last major feast for some time, and without the sugar it turns out to be a much improved one. For reasons already explained, it is wise to follow the Yorkshire habit of serving the

146

cake with a slice of hard cheese. The same cheese should as usual follow the pudding.

Christmas pudding

Do not make more than three weeks in advance of eating as its storing qualities are limited.

250g (8oz) raisins
125g (4oz) currants
125g (4oz) sultanas
50g (2oz) dried bananas, finely chopped
50g (2oz) dates, finely chopped
250g (8oz) shredded suet or Suenut
125g (4oz) brown breadcrumbs

50g (2oz) wholewheat flour
50g (2oz) grated carrot
rind of ½ lemon and ½ orange, grated or finely chopped
½ tsp grated nutmeg
½ tsp cinnamon
2 eggs
15cl (¼ pint) milk
1 tbsp brandy

Wash and drain the fruit. Mix together all ingredients except the eggs, milk and brandy. Beat the eggs. Add the milk and brandy. Mix with the other ingredients to a soft dropping consistency, adding more milk if necessary. Butter a pudding basin and almost fill with the mixture. Cover with greased paper, tie with a cloth and steam for 5 hours. Store in a cool place until required. Steam for 2 hours before serving.

Christmas cake
(to make 1.75kg/3¾lb)

35cl (12fl oz) unsweetened orange juice
2 tsp dried yeast
2 eating apples, cored
2 eggs
12.5cl (4fl oz) sunflower oil
4 tbsp skimmed milk powder
2 tbsp brandy (optional)
90g (3oz) wholewheat flour
90g (3oz) strong white flour
50g (2oz) rice flour
25g (1oz) wheat germ
25g (1oz) soya flour
50g (2oz) wheat flakes
50g (2oz) rolled oats
25g (1oz) millet flakes
90g (3oz) finely grated carrot
25g (1oz) desiccated coconut
50g (2oz) ground almonds
50g (2oz) broken walnuts
50g (2oz) broken cashew nuts
300g (10oz) mixed dried fruits, washed and drained
grated rind of 1 orange and 1 lemon
2 tsp mixed spice
½ tsp cardamom powder
½ tsp cumin powder

Warm ¼ litre (8 fl oz) orange juice, reserving the remainder for later. Disperse the yeast in the orange juice for 10 minutes, then add apples, eggs, oil and skimmed milk powder. Liquidize and set aside.

Mix all the remaining ingredients. Stir in the yeast mixture and blend thoroughly. Cover and leave to rise in a warm place for at least 2 hours or overnight if possible.

Mix in the remaining orange juice and brandy. Lightly oil and line a 20cm (8 inch) square cake tin and spoon in the mixture. Cover and leave to rise for a further 1 hour. Cover the cake with greaseproof paper and bake at 190°C/375°F (gas mark 5) for 1 hour and for a further hour at 170°C/325°F (gas mark 3). Test with a knitting needle. If it comes out clean, the cake is done. If it does not, continue baking for 15 to 20 minutes. Turn out on to a wire rack and cool completely before cutting. Store in an airtight container.

Keeping qualities are limited, so it is best made only a week in advance.

Mincemeat
(makes 1.7kg (3lb 5oz)

250g (8oz) each currants,
   sultanas and raisins
50g (2oz) each dates and dried
   bananas
250g (8oz) grated apple
250g (8oz) grated carrot
50g (2oz) finely chopped
   almonds

125g (4oz) shredded suet or
   Suenut
grated rind and juice of 1
   lemon
grated rind of 1 orange
1 tsp nutmeg
2 tsp cinnamon
2 tbsp brandy

Wash and drain currants, sultanas and raisins. Chop dates and dried bananas roughly. Mince all the dried fruit. Mix in the remaining ingredients. Cover and leave in a cool place for 24 hours. Stir throughly and store in sealed sterile jars. Keep in the refrigerator or very cool larder.

In Anglo-Saxon times the last culinary fling of the dying year was the scrambling together of the last bit of flour, thin milk and anaemic eggs, before the real fast set in. We still maintain the tradition of pancakes on Shrove Tuesday and we are joined by Russian tradition, where the pancake forms the centrepiece of many meals during 'Butter Week'. It is a yeasted pancake (blin) and hence much lighter than the English variety, allowing a wide variety of filling without producing an unacceptable heaviness. Moreover, the blin is never sweetened. The blin, and its more solid English cousin, can include sour cream or plain yoghurt, as well as a variety of vegetable, meat and fish mixtures. Furthermore, our home product can easily incorporate other ingredients in the basic mixture itself and be given whatever exotic title comes to mind.

Basic pancake recipe

90g (3oz) strong white flour
50g (2oz) wholewheat flour

1 egg
30cl (½ pint) skimmed milk

Sift the flours into a bowl. Toss in left-over bran from sieve. Break the egg into a well in the flour. Beat until all the flour is incorporated and the mixture is smooth. Gradually add the milk, beating thoroughly. Cover and leave to stand for half an hour before making the pancakes.

Or liquidize all the ingredients and allow to stand for half an hour before making the pancakes.

Fry a thin layer of batter in very hot oil until both sides are golden. Stack on a warm plate and serve with a squeeze of lemon juice or filled with stewed fruits.

French pancakes

125g (4oz) plain flour	1 egg
25g (1oz) wholewheat flour	30cl (½ pint) skimmed milk
½ tsp grated nutmeg	25g (1oz) butter

Sift the flours and nutmeg into a bowl. Toss in the left-over bran from the sieve. Break the egg into a well in the flour. Beat until all the flour is incorporated and the mixture is smooth. Heat the milk and butter until the latter has melted. Gradually add to the batter, beating thoroughly. Cover and leave for 30 minutes before making the pancakes. Fry in very hot oil on both sides. Stack on a warm plate and serve with a squeeze of lemon juice.

Pancakes with carrot

basic pancake recipe	2 tbsp sunflower oil
250g (8oz) finely grated carrots	2 tsp cinnamon powder
	2 tbsp water

Simmer the carrots with the oil, cinnamon and water for 15 minutes. Blend with the pancake mixture and fry in the usual way. Serve with a squeeze of lemon juice.

Pancakes with cabbage

basic pancake recipe
250g (8oz) finely grated
  cabbage

pepper
2 tbsp sunflower oil

Simmer the cabbage with oil and pepper in a covered pan for 20 minutes. Blend with the pancake mixture and fry in the usual way. Serve with lemon juice.

Buckwheat pancakes
(10–12 pancakes)

90g (3oz) buckwheat flour
50g (2oz) strong white flour
1 tsp baking powder
½ tsp grated nutmeg

1 egg
¼ litre (8fl oz) skimmed milk
4 tbsp cold water

Sift the flours into a bowl with the baking powder and nutmeg. Make a well in the centre and break the egg in with a little milk. Gradually incorporate the flour and beat until the mixture is smooth. Add the remaining milk and water and beat vigorously. Cover and leave to stand for half an hour before cooking. Fry in a very little hot oil until both sides are bubbly and lightly browned. Stuff with savoury fillings and serve with yoghurt or skimmed milk soft cheese.

Blini (yeasted pancakes)

1 tsp dried yeast
30cl (½ pint) warm water
250g (8oz) strong white flour
1 egg

½ tsp salt
1 tbsp sunflower oil
30cl (½ pint) warm skimmed
  milk

Dissolve the yeast in the warm water in a warm mixing bowl. Stir in half the flour. Cover and leave to rise in a warm place for about an hour. Separate the yolk from the egg white. Add the yolk and oil to the risen dough, mixing thoroughly. Fold in the remaining flour and beat until the batter is smooth. Gradually add the warm milk, cover and leave for 30 minutes.

Stir to settle and fold in the beaten egg white. Cover and leave for a further 15 minutes.

The best results are achieved if the pancakes are fried immediately after the third rising. Lightly oil a small frying pan. When hot, thinly cover the pan with batter and fry on a high heat until both sides are golden.

Fillings are traditionally savoury but blini are equally attractive served with fruit purée or bananas mashed with lemon juice, or sprinkled with cinnamon and melted butter.

SAVOURY FILLINGS FOR PANCAKES

Pancakes of whatever kind are a good vehicle for using up leftover shepherds' pie, moussaka, bolognese sauce, vegetable fry-up and so on. The field for the fillings is wide and we merely offer a few pointers.

Chicken liver filling

1 small chopped onion
oil for frying

250g (8oz) chopped chicken
    livers
pepper

Fry the onion in oil until soft. Add the livers and toss about for 5 minutes over a high heat. Season and serve a spoonful inside each pancake.

Mushroom filling

250g (8oz) chopped
    mushrooms
1 medium chopped onion

oil for frying
pepper

Fry the onion in oil until soft. Add the mushrooms and fry on a high heat for 2 or 3 minutes, then reduce for a further 10 minutes. Season and serve a spoonful inside each pancake.

Chicken and mushroom filling
(for 6 pancakes)

1 chicken quarter
1 tbsp olive oil
Black pepper
½ tsp tarragon
1 clove garlic

1 medium carrot
1 medium onion
50g (2oz) mushrooms
1 tbsp wholewheat flour
30cl (½ pint) stock

Choose a pan with a lid and brown the chicken on all sides in half the olive oil. Season with pepper, tarragon and crushed garlic. Cover the pan tightly and cook slowly for a further 30 minutes until the chicken is tender. Finely chop and fry the carrot and onion in the remaining oil until the onion is transparent. Add chopped mushrooms and continue cooking for a further 15 minutes. Stir in the flour and cook for 5 minutes. Add stock and bring to the boil stirring continuously. Chop the cooked chicken into cubes and add to the mixture. Adjust with stock if the mixture is too thick.

Savoury fish filling
(for 6 pancakes)

50g (2oz) brown rice or
   wholewheat grains
30cl (½ pint) vegetable stock
1 medium green pepper
2 medium leeks
1 stick celery
1 tbsp cooking oil
1 clove garlic

2 tomatoes
1 tsp oregano or basil
125g (4oz) fresh fish (cod,
   haddock or coley)
50g (2oz) smoked fish
Black pepper
Bunch of watercress

Wash the rice or wheat and cook in stock in a covered pan for 30–40 minutes. Add more stock if the pan begins to boil dry.

Chop and fry together the pepper, leeks and celery in oil. Add crushed garlic and cook for a minute or two. Mix in chopped tomatoes and herbs and continue frying for 10 minutes. Cut the fish into 2.5cm (1 inch) cubes and add to the mixture with a little extra stock. Cook gently for 5 minutes before stirring in cooked rice or wheat. Season with pepper. Garnish with watercress.

Filled pancakes gratinées

Arrange pancakes with a variety of fillings in a shallow fireproof dish. Cover with a generous helping of well seasoned cheese sauce. Brown under the grill for 10 minutes before serving.

SWEET FILLINGS FOR PANCAKES

Banana and pecan nut filling
(serves 1)

½ banana
¼ tsp cinnamon
1 tbsp lemon juice (or orange)

1 tbsp natural low fat yoghurt
1 tbsp chopped pecan nuts (or walnuts)

Skin and slice the banana lengthways. Place in a shallow ovenproof dish and sprinkle with cinnamon and lemon or orange juice. Bake for 20 minutes at 190°C/375°F (gas mark 5). Roll up in a pancake and serve garnished with yoghurt and nuts.

Apple and raisin filling
(serves 6)

500g (1lb) cooking apples
90g (3oz) raisins
grated rind and juice ½ orange and ½ lemon

1 tsp cinnamon powder
50g (2oz) ground almonds
Extra lemon juice to serve

Peel, core and chop the apples. Stew with raisins, rind, juice and cinnamon in enough water to cover the bottom of the pan. When the apples are tender, fold in ground almonds. Fill the pancakes and sprinkle with lemon juice before serving.

Hot fruit salad filling

1 orange peeled and chopped
1 eating apple, cored and
   chopped
1 sliced banana
1 large pear, cored and
   chopped
½oz dates, chopped
1 tsp ground ginger
1 whole clove

grated rind and juice 1 orange
grated rind and juice ½ lemon
2 tbsp sherry

Garnish
Lemon juice
yoghurt
chopped walnuts

Marinate the fruits, spices, rind and juice with the sherry for about an hour. Fifteen minutes before serving, heat in a heavy pan and simmer gently for ten minutes. Fill the pancakes and garnish with lemon juice, yoghurt and chopped walnuts.

The word 'Easter' derives from the Old English 'Eostre', formerly the goddess associated with the Spring equinox. It is a time when the earth is warming up, the first productive green shoots appearing and the hens either beginning to reproduce or to lay their sterile eggs more abundantly for our human purposes. The mythological tie-ups are obvious and the accruing symbolism of resurrection, new life and so on, fairly explicit. It is certainly a time when the egg symbolism is done to death by the sugar men. Given that the sugar culture and its high technology have been long on the scene we can only reply that the charisma is costly. The cost per ounce of chocolate is between two and three times as much when it is sculptured into eggs, chicks and rabbits. Chocolate itself is at the safer end of the spectrum of sweet danger, so perhaps we ought not to fuss too powerfully about a single bonanza, provided it stays single and unprolonged. Some emphasis should, however, return to the traditional egg, with more appropriate and less expensive options made more of.

    The tradition of painting a large number of eggs on Easter Eve and hiding them all over house and garden for children to search out in the morning is in keeping with the spirit of magic necessary to the feast. Much can also be made of what is available commercially to make Easter a funtime and not merely a poorly observed and over-sugared religious festival. The various motifs of egg,

egg-cup, chicken and rabbit are available in a wide range of non-edible forms such as egg money-boxes, wooden painted eggs and chicks, fluffy chick and egg toys, pram toy chicks, and chick-shaped candles. There is even a soap egg. The list grows year by year.

Large decorated cardboard egg shells can be used as containers for the smaller gifts, thus acting as Easter's equivalent to the Christmas stocking. Exceptionally they can hold also perhaps a stick of liquorice and a few diabetic chocolate drops.

In the last few years Boots has started a line in diabetic chocolate Father Christmases, eggs, and rabbits in which the sweetener is sorbitol. A few hours before the festival close-down these are often halved in price.

In the kitchen, Easter starts with the Good Friday hot cross bun, whose symbol actually predates Christ. The sugar here gets into the spice and occasionally the cross through the use of candied peel. It is the spice and the cross that are all important, however, and sugar again is proved irrelevant.

Hot cross buns

1 tsp dried yeast	2 tsp mixed spice
2 tbsp warm water	½ tsp cumin powder
20cl (7fl oz) skimmed milk	4 tbsp sunflower oil
1 beaten egg	125g (4oz) currants/sultanas
500g (1lb) strong white flour	(washed)
1 tsp salt	50g (1oz) finely chopped dates

Dissolve the yeast in the warm water. Warm the milk and add this and the egg to the yeast. Sift the flour, salt and spices into a bowl. Rub in the oil and stir in the dried fruit. Make a well in the centre. Pour in the yeast mixture and mix to a slack dough, adding a little milk if necessary. Cover with a cloth and leave to rise in a warm place for 2 hours.

Knock the dough down and turn on to a floured board. Divide and shape into 12 buns. Cut a cross on each with a sharp knife. Brush with milk. Cover and allow to rise for a further 15 minutes. Bake for 25 minutes at 220°C/425°F (gas mark 7). Cool on a wire rack.

The tradition of a special cake for Easter is not strong in Britain. The baking of a simnel cake, first for Mothering Sunday and later for Easter, is a lost art which we are happy to revive in a less contaminated form. Another possibility is a further Russian borrowing, the yeast-raised *kulich* which has the added advantage of being small and allowing total family involvement. Our unsugared version is of a cake that in Russia is traditionally over-sweet.

Easter cake

With their bundles and *kuliches* the people pressed close to each other among the crosses and gravestones. Most of them had evidently come a long way to have their *kuliches* blessed.

Anton Chekhov
Easter Eve

1 ¼ tsp saffron
   6 tbsp vodka, white rum or
      brandy
   2 tsp dried yeast
   4 tbsp warm water

   40cl (13fl oz) warm skimmed
      milk
   350g (12oz) strong white flour
   125g (4oz) wholewheat flour

2 15cl (¼ pint) sunflower oil
   2 eggs
   ½ tsp cardamom powder

   175g (6oz) strong white flour
   50g (2oz) wholewheat flour

3 175g (6oz) raisins
   25g (1oz) chopped dried
      banana
   90g (3oz) chopped dates

   50g (2oz) almonds, roughly
      ground or chopped with
      skins on
   grated zest of 1 orange and 1
      lemon

1 Soak saffron in vodka for about 2 hours or overnight. Dissolve the yeast in the warm water and add this and the milk to the mixed flours in a bowl. Mix thoroughly, cover with a cloth and stand in a warm place to rise for about 45 minutes by which time the dough should have doubled its size.

2 Separate the yolks from the egg whites. Add yolks, oil and cardamom powder to the risen dough and mix well. Blend in the remaining flours and then fold in the stiffly beaten egg whites.

Cover and leave to rise in a warm place for a further 45 minutes.

3 Place the remaining ingredients in a separate bowl. Strain the saffron infusion over the top and mix well. Add the whole lot to the twice-risen dough and blend thoroughly.

Cut and lightly oil 12 strips of greaseproof paper measuring 17.5×12.5cm (7×5 inches) and 24 rounds 5cm (2 inches in diameter. Oil 12 patty tins and arrange a 'chimney' in each. Fill one-third full with dough and place a round of paper on top of each to prevent burning. Oil and line a 15cm (6 inch) diameter cake tin and fill with the remaining dough, placing a circle of paper on top. Bake the little cakes for 45 minutes at 180°C/350°F (gas mark 4) and the large cake for about 1 hour. Remove from the tins and cool on a wire rack. When quite cold, store in an airtight container.

If they last long enough to become stale, slice, toast and butter.

Simnel cake

1 tsp dried yeast
2 tbsp warm water
4–6 tbsp skimmed milk
4 tbsp sunflower oil
2 eggs
1 tbsp brandy (optional)
250g (8oz) strong white flour
50g (2oz) currants

40g (1½oz) sultanas and/or
  raisins
25g (1oz) finely chopped dates
grated rind of ½ lemon and ½
  orange
1 tsp powdered cinnamon
½ tsp grated nutmeg

Almond paste
90g (3oz) almonds
3 tbsp oil
15g (½oz) dates
juice of 1 orange

2 tsp orange flower water
  (obtainable from
  pharmacists)

Disperse the yeast in the warm water for 10 minutes or so. Warm the milk to no more than blood heat and add to the yeast mixture with the oil, beaten eggs and brandy. Sift the flour into a mixing bowl. Make a well in the centre and pour in the yeast mixture. Gradually incorporate the flour until all is thoroughly blended. Cover and leave to rise in a warm place for about 2 hours.

Wash the dried fruit and mix with the chopped dates and rind.

Coat with cinnamon and nutmeg and fold into the risen dough. Cover and set aside. Grease and line a 15cm (6 inch) cake tin.

To make the almond paste, grind almonds with the skins on, adding the oil gradually. Soften the dates by warming gently in the juice of an orange. Liquidize with the almonds and add the orange flower water.

Put half the dough into the prepared cake tin, spread on the almond paste and cover with the remaining dough. Protect the top of the cake with a circle of greaseproof paper and allow to rise for a further 20 minutes before baking. Heat the oven to 180°C/350°F (gas mark 4) and bake the cake for 1 hour. Test with a knitting needle. Remove from the tin and cool on a wire rack.

If the cake lasts long enough to become stale, slice, toast and butter.

Our autum feastings have become something of a muddle. Historically it was definite shifts in climate which determined the beginnings and endings of seasons. Thus the Celts declared the beginning of winter in early November when the cattle were brought down from pastures. Their festival of Samhain on November 1st is now overlaid by All Saints, All Souls (Hallowe'en), Gunpowder Plot, and Thanksgiving. So around the bonfires it is scarves, wellies and woollies, a crackling of baked potatoes, bangers and beefburgers whose fat we hope will drain off sufficiently to keep up heart and spirit. Not much of a look in for our old enemy except through the pumpkin, a sweet enough old vegetable much underused. We find it sad that North Americans should celebrate the safe landing of their first Devonian settlers with a pie made from a sugared supermarket cake-mix. The following recipe should make it more of a culinary event.

Pumpkin tart

(two 25cm (10 inch) tarts)

500g (1lb) low fat short pastry (see page 119)
15cl (¼ pint) milk
2.5cm (1 inch) piece vanilla pod (or ½ tsp vanilla essence)
500g (1lb) pumpkin (flesh weight)

50g (2oz) wholewheat breadcrumbs
40g (1½oz) chopped nuts
50g (2oz) chopped raisins
½ tsp ginger powder
¼ tsp ground cloves
1 tsp cinnamon powder
Juice and zest ½ orange and ½ lemon

Lightly oil and line with short pastry two 25cm (10 inch) foil cases and refrigerate until required.

Warm the milk in a pan with the vanilla pod (or essence). Set aside while preparing the pumpkin. Cut the flesh into largish cubes and cook in a tightly covered pan for about 15 minutes, with enough boiling water to cover the bottom of the pan. Drain (reserving stock for soup) and set aside.

Remove vanilla pod and soak the breadcrumbs in milk. Chop the nuts and raisins and put into a mixing bowl with the spices, orange and lemon juice and zest. Separate egg yolks from whites. Liquidize the yolks with the pumpkin and add this and the soaked breadcrumbs to the nuts and raisins. Mix thoroughly, then fold in the stiffly beaten egg whites.

Divide the mixture between the two pastry cases and bake at 220°C/425°F (gas mark 7) for 15 minutes. Reduce to 180°C/350°F (gas mark 4) for a further 30 minutes. The finished tart is slightly risen and cracked on top. Cool on a wire rack before serving.

There are few traditions to rely on for that hardy and unpredictable annual, the child's birthday party. Magic, excitement and style are all important here, but no recipe for success is sure and pitfalls abound. The details of the required magic are seldom obvious and a perverseness of response can ruin every best laid plan. There is often a wide gap between what parents think will go down well, perhaps what they like to make, and what wins on the day.

The prospect of, say, apple pie might give rise to a chorus of delight from children who will prevent its production by a constant clamour for pieces of raw pastry and flour-dredged but unsugared raw apple. No mother-proud featherlight cake ever reaches the charismatic heights of a scrape round its mixing bowl. The parent who has toiled in imagination and sweat to make the birthday cake in the shape of a battleship, will be rewarded with gasps of wonder and delight from children who will then go in search of the crisps and beefburgers.

Sugar can no more guarantee success than anything else and no easy solutions are on offer. For, though when children go to a party they expect something to eat, stomach satisfaction is not the central theme. Only when imagination is lacking on the entertainment and games side will the food come under close critical scrutiny. We suggest therefore that, for the sake of energy conservation, the food should be arranged as follows.

The centre piece need not necessarily be a cake. If it is, then the obvious choice is the spicy fruit cake in Chapter 9. It can be spread with a thin coating of cocoa cashew spread (page 113) or almond paste (above) sufficient to glue on coloured or plain desiccated coconut in patterns or words.

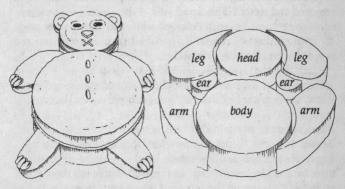

Alternatively the chocolate cake (page 102) can be made into a teddy bear. Slice it in half horizontally and cut into two small and two large matching rounds. The remaining crescent shapes will make up arms, legs and ears. Spread one side of each matching shape with cocoa cashew spread (page 113), glue shapes together and assemble the teddy on a tray, board or large plate. Paste the top either with the spread or cream and decorate with desiccated coconut, using nuts, currants or fresh fruit for eyes, nose, mouth and buttons. Surround with coconut.

Two summer alternatives involve the ice cream and jellies on pages 128–130. For the ices version, prepare lots of seasonal fresh fruit cut into smallish pieces as for a fruit salad, adding a few nuts and raisins for variety. Just before the display and eating, scoop the ice cream from its container making a wide ring on a very cold dish, and fill with the fruit, the centre piece being a half orange or grapefruit to carry candles. The circular mould of an ice cream maker will make life easier for this one.

Jellies can be moulded and surrounded by pieces of cored apple and orange segments stuck with candles.

Of the fun snacks, the gingerbread men and the Johnny cake are the least fussy to prepare, and contribute a frisson of excitement associated with mock cannibalism and escape from the fiery furnace.

Slightly more fiddly are hard boiled eggs in animal shapes. Each should be shelled and cut in half lengthways and the yolk halves removed and sieved. The sieved yolk is then mixed with mayonnaise and returned to each hollowed white and animal faces made up from cuttings of colourful vegetables such as green and red peppers, olive, carrot and cucumber.

After this the food provider had better stick to the basics of sausages on sticks, sausage rolls, cheese straws, cheese-and-grape sticks, salt or water biscuits with cottage cheese dip and burgers.

The absence of sugar in drinks is not as restrictive as it seems. Besides teas and tisanes of many kinds, milk, Bovril, Marmite, unsweetened juices and low-calorie drinks (listed in Appendix 1) there are many interesting and exotic combinations to add excitement. Many of the following ideas stem from the Indian tradition of serving yoghurt-based drinks, particularly refreshing on hot summer days. For the flagging adult, these can be suitably laced as the two-hour stint threatens to seem endless.

Strawberry yoghurt drink

250g (8oz) strawberries
15cl (¼ pint) natural low fat
  yoghurt
15cl (¼ pint) cold water

2 tsps lemon juice
1 tsp orange flower water
  (optional)

Liquidize the strawberries then add the remaining ingredients.
Beat vigorously. Chill and serve with ice, lemon slices and mint
leaves.

Banana and redcurrant yoghurt drink

125g (4oz) redcurrants
250g (8oz) banana (skinned)
15cl (¼ pint) natural low fat
  yoghurt

30cl (½ pint) cold water
1 tsp rose water (optional)
1 tsp lemon juice

Liquidize and strain the redcurrants. Blend with the remaining
ingredients, beating vigorously. Chill and serve with ice, slices of
orange and cucumber or mint leaves.

Cucumber drink

125g (4oz) cucumber
1 tbsp fresh dill
1 tbsp fresh lovage leaves
2 tsps lemon juice

15cl (¼ pint) natural low fat
  yoghurt
30cl (½ pint) cold water

Liquidize all the ingredients. Strain and chill. Serve with slices of
orange, ice and mint leaves.

Prune and yoghurt drink

125g (4oz) prunes
2.5cm (1 inch) stick cinnamon
30cl (½ pint) water

15cl (¼ pint) natural low fat
  yoghurt
1 tbsp lemon juice

Soak the prunes overnight with the cinnamon and water. Bring to the boil and cook for 20 minutes. Cool and discard the stones and cinnamon. Liquidize with the juice, adding yoghurt and lemon juice and extra water to taste. Strain, chill and serve with sliced lemon and ice.

Orange apricot cup

50–90g (2–3oz) dried apricots
30cl (½ pint) water

30cl (½ pint) unsweetened
orange juice

Soak the apricots for about an hour in the water. Bring to the boil and cook for 10–15 minutes. Cool, liquidize and strain, adding the orange juice and extra water to taste. Chill and serve with sliced lemon and ice.

Orange milk shake

30cl (½ pint) unsweetened
orange juice

4 tbsp skimmed milk powder

Liquidize or whisk vigorously and serve with slices of orange.

Fruit punch

60cl (1 pint) unsweetened
orange juice or pineapple
juice
60cl (1 pint) unsweetened
apple juice

30cl (½ pint) unsweetened
grapefruit juice
juice of ½ lemon
pieces of chopped orange,
lemon, apple and mint
leaves to garnish

Mix all the ingredients. Garnish with fruits and mint and serve chilled with crushed ice.

Mulled fruit punch

60cl (1 pint) unsweetened
   orange juice or pineapple
   juice
60cl (1 pint) unsweetened
   apple juice
30cl (½ pint) unsweetened
   grape juice
juice of 1 lemon

1 chopped apple
1 chopped orange

Spices
½ tsp ground ginger
2.5cm (1 inch) cinnamon stick
¼ tsp nutmeg
2 whole cloves

Mull all the ingredients over gentle heat for about 20 minutes and serve.

The final assault by the sugar culture on the birthday feast is in the parting gift. Chocolate bars, lollies and tubes of sweets lie in wait at the supermarket exit for the exhausted shopper, who has only to transfer them to the domestic check-out point for the feast to be complete. Taste-buds already stripped of sensibility and stomachs grappling with satiety are faced with new and irrelevant burdens.

The answer is, of course, to leave the calories on the table where they belong and round off the event with gifts that prolong the absorption and fun. A pencil and notebook, a balloon, a colouring book and so on, will give mouth and stomach a rest and minds a final burst of activity at the end of the long day. With such little effort the birthday feast, like all the rest, need admit no skeleton.

13 Coming clean
Sense and nonsense in toothcleaning

In the hierarchy of magical practitioners, and below the medicine men in prestige, are specialists whose designation is best translated 'holy-mouth-men!' The Nacirema have an almost pathological horror of and fascination with the mouth, the condition of which is believed to have a supernatural influence on all social relationships. Were it not for the rituals of the mouth they believe that their teeth would fall out, their gums bleed, their jaws shrink, their friends desert them, and their lovers reject them.

. . . this rite involves a practice which strikes the uninitiated stranger as revolting. It was reported to me that . the ritual consists of inserting a small bundle of hog hairs into the mouth, along with certain magical powders, and then moving the bundle in a highly formalised series of gestures.

Horace Miner
'Body Ritual among the Nacirema'
in *American Anthropologist*, June 1956

Years ago I was travelling through Natal on horseback, and I was anxious to find a lodging for the night, when I came across a hut evidently occupied by a white man, but nobody was about.

In looking around inside the hut, I noticed that although it was very roughly furnished, there were several toothbrushes on what served as a wash-hand stand so I guessed that the owner must be a decent fellow, and I made myself at home until he came in, and I found that I had guessed aright.

Sir Robert Baden-Powell
Scouting for Boys

Many cultures accord magical significance to teeth. As they are the last part of the body to disintegrate after death, the wearing of tooth necklaces in some Pacific cultures has been thought to promote longevity. As a prelude to his sexual essays the pre-pubertal Dobu male blackens his teeth with paint[1]. In parts of Vietnam and Thailand[2] the teeth of women are black-lacquered because white teeth are considered unsightly. Habits of tooth-filing and other dental mutilations are widespread in Africa and Asia[3].

While some peoples make attempts at cleaning with bamboo or ivory picks, it is only in Western 'civilized' societies that a tooth-cleaning ritual enjoys social prestige. Here it is one of the indices of respectability, separating the genteel from the vulgar, the nice from the nasty, the righteous perhaps from the sinful. The white-toothed smiles of North American show business, from Doris Day to the Osmonds, seem to offer a paradigm of rectitude and wholesomeness to which we must all aspire through cleanliness. Indeed our attitudes to the mouth's inherent frailty have always had more to do with muscular Christianity, and a yearning for proximity to godliness, than the actualities of health and disease. Until two or three decades ago little was precisely understood about the nature of this frailty and what exactly could be done about it, so that the advised cleaning procedures had no more basis in rationality then the rituals of South East Asia. Thus the formal gum-to-tooth brushing exposed in Chapter 1 was prescribed as an antidote to the punishing flagellations of the obsessive which often seriously damaged teeth by abrasion. The preventive element here was directed against trauma, not disease.

The boom in fundamental research that took place in the early sixties removed whatever scientific coherence these rituals enjoyed. What was now established was that the sugar/plaque interaction was so rapid as to render the after-meal dash to the washbasin an ineffectual contributor to dental health. We learned also that plaque behaves independently of food in that when removed it re-forms in between 36 and 48 hours whether food is taken or not. This enabled the preventionists to concentrate on the plaque and forget about the food. We could separate tooth-cleaning from meals and simply shift the plaque once a day at our convenience. The level of cleanliness required was very high and to achieve it Preventive Dentistry offered hours of 'motivation', instruction, checking and cleaning in the dental surgery. Calls came from the high priests to set up 'Oral Hygiene Centres' under the NHS to dispense the new blessing.

Both the abandonment of the after-meal habit and the replacement of the gum-to-tooth ritual by more rationally based cleaning methods are the twin achievements of the preventionists.

This, however, leaves the toothbrushing layman in various sorts of difficulty because unfortunately the old advice is still left lying around. It is in fact pure chance whether he hits upon sensible advice or not. Dentists and their manufacturing allies who fund most advice drop not the ghost of a hint that the advice is *new* or has ever been different. In *Caring for Teeth*, the Consumers' Association recommended the gum-to-tooth method but this was quietly dropped in the April 1980 issue of *Which*? In this *same* issue it recommended toothpaste which it had found valueless in 1975. Coming clean is apparently OK for teeth but not for advisers on them.

The unceasing and incoherent din of dental health advice provoked the Health Education Council in 1978 to make the understatement of the year.

> It would seem that the information presented to the public at
> present by dental health educators is unnecessarily complicated,
> frequently contradictory, and sometimes wrong.[4]

Until this very untidy house is put in order we ought to bear in mind that such a simple matter as cleaning your teeth ought, of course, never to be cluttered with the paraphernalia of the professional mind or the lunatic proliferation of aids and motivational apparatus, turning a simple matter into a recondite series of mysteries, suitable only for the confessional of dental surgeries. So where to start?

First of all we have to ask, why should we do it? The question is rarely put. There is really only one scientific reason. If we do it effectively we can stop gum bleeding and its associated disease. In spite of claims for decay prevention by plaque-control there is not a lot in the way of hard evidence to permit such optimism. As we have shown, decay prevention is a matter of getting the sugar levels drastically down, of not violating the mouth's ecology.

So without aiming for the bacterial no-go area dear to the

ambitions of the plaque-controllers, we can accept much of their case and get the plaque levels down also. As plaque is pale it has to be 'disclosed'. This simply means dabbling a food colour on to the teeth, with a cotton bud perhaps, rinsing and seeing where the stain has held. There are expensive lozenges which are sold for the purpose at pharmacists. Not only are they no more effective than the food colourant,[5] there's something odd about dealing with a sugar-based disease by giving people a sweet to suck!

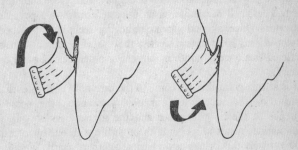

How to miss it . . . how to shift it

Whatever is used, the important thing achieved by the staining is that it liberates us from notions of correctness-via-ritual. Given good visual access by use of a cheap mouth mirror (now available from pharmacies), and a shaving mirror with adjacent table lamp, we can work it out for ourselves. If the method shifts the stain, we are doing it right, if it doesn't, we are not. Simple scrubbing with a brush angled a little 'towards' the gum is as good a method as any. The choice of brush need not be agonizing. It should, for prefer-ence, be of medium hardness, rather small (child size) and thickly tufted. The smallness is merely to prevent retching when the back teeth are brushed. The actual brushing need only go on until the stain has gone. Any bleeding provoked should not cause alarm, since given two or three days of efficient tooth cleaning the gum will heal, having no longer the proximity of the ulcerating plaque.

None of this requires toothpaste, which merely gets in the way of vision. Furthermore it is *sweet* and counterproductive to the low sweetness strategy central to dental health. The minty flavours also give a false impression of cleanliness even when plaque is present.

We have to say in its favour that it does help to remove stain, people like it and fluoride has given it the first genuine decay-preventing ingredient. There remains however, a clear and unresolved opposition of aims in using a paste to *remove* something *from* teeth, while it contains a substance we want to *put into* them.

Given that paste-use is what is called a 'sociological norm', however, we suggest a compromise. If previous tooth cleaning procedures have been meal-oriented and plaque-control is being attempted for the first time, then a holiday from paste should be taken. The smell of plaque on an unpasted brush is a powerful motivator. Once a good plaque-removing technique has been developed, say after a few weeks, the use of colourant can be restricted as a once-a-month monitor. For other times a paste can be used.

It is at this point that we come to the threshold of the dentist's world. For the success of our dietary changes, the possible influence of the fluoride paste and the mystery factor will, in combination, result in us going through life with our own teeth. Unfortunately this brings us up against another hazard. We know that in the over-35 age group the chief cause of tooth loss is not decay but gum disease. As more of us take more of our teeth healthily into our thirties, forties and beyond, the level of gum disease in the population may well rise to epidemic proportions. Our local neighbourhood dentist is going to be less a tooth-repairer and more a gum specialist and, though the simple brushing described above will deal with most early gum disease, more specialized approaches will be required for many. While the use of other devices for more sophisticated cleaning can, given the proliferation of aids now available, remain a DIY matter it is possibly quicker and safer to be tutored by a good hygienist and the choice of dentist may be more and more governed by the quality of care offered in this field.

Most of us will have no need or time for the complexity of gadgetry and the obsessive level of commitment to mouth scourging demanded by the manufacturers and the plaque-besotted dentists. Anyone who wants clean teeth can disclose with a colourant and use a toothbrush. It doesn't require a university degree.

Within the dentist's world there are two other 'preventive' expedients. The oldest is the so-called topical application of fluoride (not funded by the NHS) to harden tooth enamel. This has been critized for low cost-effectiveness in the past and seems to be of even more limited value now that fluoride pastes are taking 97% of the market. The other (also not NHS funded) is the sealing of the fissures of young back teeth by means of a thin plastic coat. Such fissures (i.e. the grooves on the chewing surfaces) are the site of the earliest attacks of decay. While preventive theologians argue about whether it constitutes 'true' prevention, it is, from the child's point of view, much more congenial than the drill, and the results are far less unsightly than fillings.

Useful as these expedients are, they do not amount to a powerful armoury. Difficult though it is for dentists, with all the highly sophisticated panoply of skills and practical experience they can now deploy, to concede the point, the key to dental health is not in their hands. Difficult too for lifelong patients to accept that it is not available by plugging into any treatment system, especially when they are constantly being told that it is.

We have, therefore, to come back to base and remind ourselves of the real issues. The hard and inescapable truth is that, the stagnant issue of water fluoridation apart, the meagre capacities of health services are dwarfed utterly by the political and commercial monstrosity of sugar economics. This pure chemical, which supplies almost one fifth of our average calorific needs without conferring any other nutritional benefit, has blasted through our lives and kitchens, destroying in its path our more sensible culinary habits, our waistlines, our health in general and our mouths in particular. Its omnipresence has so enfeebled our professional advisers that life without it seems to them unimaginable. When, therefore, they come to measures of control, their sense of what is normal is never a nutritional one, since there is no nutritional requirement for it, only a sociological one, i.e. what is habitual. The resultant advice is reduced to a pathetic 'Eat less' injunction without benefit of quantification.

So again we are on our own. Faced with the mighty clout of the sugar manufacturers and their political backers, the individual decisions required will not be easy to mobilize. This book has been designed specifically to buttress those decisions with historical perspective and contemporary information. Although the battle to keep our teeth from the predators is ultimately the larger political one, we are not helpless. We certainly cannot wait, while we have children growing up, for the larger political solution either. The previous chapters have pointed a way out of the commercial and cultural trap. It may not amount to total liberation but, in the sense that knowledge of constraint is the beginning of liberty from it, we have a good start.

Appendix 1
Recommended products without added sugar

These listings were compiled mainly from information supplied by manufacturers, for whose help we are most grateful. Where limited or no information was supplied, it was obtained by legwork around the shops. One supermarket chain felt unable to cooperate feeling that, as formulations change frequently, there is a risk that information about them soon becomes out of date. This highlights the need for constant consumer vigilance in checking ingredient panels.

A short list of Australian products appears at the end.

'Humanized' milks *See Chapter 4*

SMA, SMA Gold Cap, Cow & Gate Babymilk Plus, Cow & Gate Premium, Osterfeed.

Teething preparations *See Chapter 4*

Anbesol Flavoured Gel, Bonjela, Hewlett's Teething Jelly, Maw's Soothadent, Moore's Teething Jelly, Steadman's Teething Jelly, Oral-B Dental Gel.

Tinned fruits *See Chapters 5 & 8*

i.e. several brands of pineapple in natural juice.

Australian Gold
Cling peach halves in fruit juice, Pear halves in fruit juice.

Boots (Shapers)
Pineapple chunks, Pear quarters, Peach slices, Fruit cocktail, Apricot halves. All saccharin sweetened.

Ceramin's Natur Fruit
Pear halves in fruit juice, Peach slices in fruit juice, Apricot halves with apricot juice.

John West
Grapefruit segments, Pear quarters, Fruit cocktail, Peach slices, Raspberries in fruit juice, Guava halves in fruit juice.

Sainsbury
Range of fruits in natural fruit juices (Apricot halves are in apple juice).

Tesco
Grapefruit in natural juice, Pineapple slices in natural juice.

Rusks *See Chapter 5*

Bickiepegs.

Cereal products for babies *See Chapter 5*

Milupa
Seven Cereal, (Note: Seven Cereal Breakfast contains honey), Rice Cereal.

Farley
Farex, Farlene, Baby Rice.

Robinson's
Baby Porridge Oats, Mixed Cereal, Protein Baby Food.

Boots
Baby Porridge Oats, Protein Baby Cereal.

Cow and Gate
Baby Cereals.

Health Valley
(English distributor: Harmony Foods), Sprouted Baby Cereal, Brown Rice Baby Food.

Cereals *See Chapter 8*

Nabisco
Shredded Wheat.

Quaker
Puffed Wheat.

Sainsbury
Mini-Wheats, Puffed Wheat.

Also permitted:
Grape Nuts.
Energen Wheat Flakes.
Ready-Brek (plain).

Quaker
Warm Start.

Sainsbury
Instant Hot Cereal.

Muesli *See Chapter 8*

Sunwheel.

Cheshire Whole Foods.

Life & Health Foods (Norwich)

Familia
Swiss Birchermuesli (marked 'no added sugar').

Boots
Second Nature

Kelloggs
Summer Orchard

Sainsbury
Deluxe Muesli

Tesco
Wholewheat Muesli

Diabetic Marmalades *See Chapter 8*

Frank Cooper.

Boots.

Yoghurts

All natural low fat yoghurts.

St Ivel
Shape Fruit Yoghurts.

Chewing Gum *See Chapter 9*

Oribit.

Trident.

Crisps/Snacks *See Chapter 9*

Allinsons
Wheateats: Cheese and Pizza.

Bahlsen
Barbecue Potato and Wheat Snacks, Chipsletten, Salzletten, Salz-brezel.

Co-op
Ready Salted Crisps.

Flavor Tree
Cheddar Chips, French Onion Crisps, Sesame Chips, Sesame Party Mix, Sesame Sticks.

Freddy Bee
Corn Puffs, Crisps.

Golden Wonder
All Stars, Atom Smashers, Country Crunch, Crackles, Ready Salted Crisps, Ringos, Slightly Salted Wotsits, Royales Crinkle Potato Chips, Royales Prawn Cocktail.

Health Valley (English distributor: Harmony Foods)
Regular Potato Chips and Dip Chips, No Salt Potato Chips and Dip Chips, Corn Chips, Corn Chips with Cheese, Sesame Corn Chips, Cheese Puffs, Tortilla Strips, Yoghurt Tortilla Strips, Cheese Tortilla Strips, Vegetable and Herb Tortilla Strips, Cheese Crackers.

K.P.
Cheese & Onion Crisps, Cheesy Crunchies, Crinkly, Discos, Hula Hoops, Mini Chips, Outer Spacers, Potato Crisps (many flavours), Sky Divers, Good 'n' Crunchy.

Marks and Spencer's
Crisps: Plain, Home Style, Barbecue Beef and Onion, Prawn Cocktail, Salt and Vinegar, Cheese and Onion. Chiplets (salt and vinegar flavour), Cheese Tasters, Savoury Puffs, Potato Sticks, Wheat Crunchies, Potato Rings.

Sainsbury
Bacon, Cheese flavoured Potato Snacks, Cheese and Onion Potato Snacks, Cheesy Chunks, Crunch Sticks, Plain Crisps, Potato Chips, Potato Swirls, Prawn Cocktail Potato Snacks, Twiglets.

Smiths
Bones, Chipitos, Chipsticks, Claws, Fangs, Football Crazy, Horror Bags, Monster Munch, Potato Crisps, Quavers, Quirls, Smokees, Twists.

Spicer
Wheateats.

Walkers
Bacon Snaps, Beef and Onion Crisps, Cheese and Onion Crisps, Cheese Snaps, Ready Salted Crisps, Roast Chicken Crisps, Salt and Vinegar Crisps, Smoky Bacon Crisps, Spicy Tomato Snaps.

Biscuits/Crispbreads *See Chapter 9*

Allinson
Wholemeal Crispbread.

Askey
Wafers, Cornets.

Associated Biscuits Ltd (Huntley & Palmers, Jacobs, Peek Frean)
Bath Olivers, Cheeselets, Cornish Wafers, Cream Crackers, Rye Crispbread XXX, Twiglets, Vita Wheat, Vita Wheat Crispbread, Vita Wheat Rye, Water, Water Highbake.

Buitoni
Melba Toast.

Carr's
Table Water.

Crawfords
Butter Puffs, Cheddars, Tuc.

Energen
Brownwheat, Cheese, Rye, Starch Reduced Crispbreads, Wheat Bran Crispbread.

Ideal
Bran Thin Crispbread with Sesame Seeds, Ultra-thin Norwegian Crisp Bread, Whole Grain Norwegian Crisp Bread.

Jans
Rye Crispbread, Rye Crispbread with Linseed, Rye Crispbread with Sesame.

Krispen
Wholewheat and Bran Crispbread.

Lyons Maid
Cornets, Wafers.

McVities
Brown Rye, Ry-King, Ry-King Fibre Plus.

Nabisco
Cheese and Onion Wheatmeal Sandwich, Snack Crackers, Wheat Crackers.

Primula
Rye-Bran Extra Thin Crispbread, Rye-Bran Thick Crispbread.

Roka
Cheese Crispies.

Ryvita Co Ltd
Light Rye, Ryvita Danish Style, Ryvita Original, Ryvita Swedish Style Brown, Crackerbread.

Sainsbury
Cream Crackers, Rye Crispbread, Wheat Crispbread.

Scanda
Scanda Crisp.

Sharwood
Puppadoms.

Walls
Cornets, Wafers.

Beverages *See Chapters 8 & 9*

Oxo, Bovril, Marmite, Mineral waters

Beecham
Hunts Low Calorie American Ginger Ale, Hunts Soda Water, Hunts Unsweetened Orange Juice, Hunts Tomato Juice Cocktail, PLJ Lemon Juice (original sharp), PLJ Lemon Juice (less sharp), Idris Soda Water, Diet Tango, Sparkling Low Calorie Orange Drink, Diet Quosh Low Calorie Lemon Drink, Diet Quosh Low Calorie Orange Drink, Diet 7-Up.

Boots
Cola, Lemonade, Orangeade, Sweetened Lemon Juice, Unsweetened Lemon Juice.

Britvic
Bitter Lemon Crush, Cola, Grapefruit Crush, Orange Crush, Pineapple Juice, Slimsta American Ginger Ale, Slimsta Indian Tonic Water, Soda Water, Tomato Juice, Tomato Juice Cocktail (contains Worcestershire Sauce).

Canada Dry
Slim Bitter Lemon, Slim Cola, Slim Ginger Ale, Slim Tonic Water, Slim Lemonade, Soda Water, Diet Cola.

Coca-Cola
Fresca, Tab.

Corona
Diet Cherryade, Diet Lemonade, Diet Orangeade, Diet 7-Up, Diet Tango Orange, Diet Coca Cola.

Energen
One-Cal Appleade, One-Cal Cola, One-Cal Lemonade, One-Cal Orangeade, One-Cal Lime and Lemonade, One-Cal Tonic Water, One-Cal American Ginger Ale, One-Cal Bitter Lemon.

Lockwoods
Bitter Lemon, Cola, Lemonade, Lime and Lemon, One-Cal Bitter Orangeade.

Peter Eckes
Drink 10.

Rawlings
Unsweetened Orange Juice, Unsweetened Pineapple Juice

Renown Products (Penge) Ltd (Renpro)
Orange Drink, Lemon Drink, Blackcurrant Flavour Cordial, Lime Juice Cordial, Lemon and Lime Drink.

Sainsbury
Pure Orange and Passionfruit Juice, Orange Juice, Apple Juice, Grapefruit Juice, Pineapple Juice, Grape and Apple Juice, UHT Jaffa Orange, UHT Jaffa Grapefruit, UHT Apple, UHT Pure English Apple, Lemonade, Cola, American Ginger Ale, Bitter Lemon, Tonic, Lemon Squash, Orange Squash, Lager.

Schweppes
Slimline Low-Cal Shandy, Slimline Low-Cal Tonic Water.

Schweppes International Ltd
Roses Diabetic Lemon Squash, Roses Diabetic Orange Squash, Slimline American Ginger Ale, Slimline Bitter Lemon, Slimline Lemonade, Slimline Lemonade Shandy, Slimline Sparkling Orange Drink, Slimline Tonic Water, Appeltise, Soda Water, Unsweetened Orange Juice, Unsweetened Pineapple Juice.

Shapers
Cola, Shandy, Bitter Lemon, Lemonade, Orange Drink, Lemon & Lime, Orangeade.

Strathmore
Trim Low Calorie American Cola, Trim Low Calorie Lemonade, Trim Low Calorie Lemon Flavour Drink, Trim Low Calorie Lime Flavour Drink, Trim Low Calorie Orange Squash.

Spreads *See Chapter 9*

Burgess
Genuine Anchovy Paste.

Country Pots
Chicken and Salami, Pork Paste with Herbs, Smoked Mackerel.

Eden Vale
Cottage Cheese: Chives, Onion and Pepper, Pineapple, Plain.

Elsenham
Gentleman's Relish.

Farquhar North & Co Ltd (Far North label)
Pastes: Beef, Beef and Ham, Beef, Ham and Tomato, Bloater, Chicken, Chicken and Ham, Crab, Kipper, Liver and Bacon, Pilchard and Tomato, Sardine and Tomato, Salmon and Shrimp, Tongue and Ham, Tongue and Turkey. Spreads: Beef, Chicken, Salmon. Potted Salmon.

Philadelphia
Full Fat Soft Cheese: plain, garlic and herbs, chopped chives.

Princes
Pastes: Beef and Bacon, Beef and Onion, Chicken and Ham, Liver and Bacon, Mackerel and Cucumber, Salmon and Cucumber. Pâtés: Pâté de Campagne, Pâté de Foie de Vollaille, Pâté de Venaison.

Sainsbury
Chicken Spread, Minced Chicken, Salmon and Anchovy Paste, Salmon with Butter Spread, Salmon and Shrimp Paste, Sardine and Tomato Paste, Spiced Ham Spread, Veal and Ham Paste.

St Ivel
Cottage Cheese: Natural Onion and Chives, Onion and Pepper, Pineapple. St Ivel Cream Cheese.

Shippams
Pastes: Anchovy, Bloater, Chicken and Ham, Beef, Ham and Beef, Pilchard and Tomato, Salmon and Shrimp, Sardine and Tomato. Spreads: Chicken, Ham.

Cheese spreads

Sugar in the cheese spreads is a rarity. We can find no case of an unflavoured spread containing sugar. New varieties from all parts of the EEC enter the supermarkets by the week and any list would soon date. Ingredients panels which invite close scrutiny are those where the flavour is any kind of fruit. The more sweetness in the natural fruit, the more tends to be added.

Pickles and sauces *See Chapter 11*

Binisa Oriental Foods
Mixed Pickle in oil, Mango Pickle in oil, Chilli Pickle in oil.

Burgess
Capers in Vinegar, Genuine Anchovy Essence, Mushroom Ketchup.

Colmans
Dry Sauce Mixes: Bread, Cheese, Mushroom, Onion, Parsley, White. Dijon Mustard, Garden Mint, German Mustard, Mustard Powder.

Crosse & Blackwell
Capers, Cocktail Onions, Gherkins, Mixed Pickles, Onions (White), Stuffed Manzanilla Olives.

Goldenfry
Parsley Sauce Mix, Rich Brown Gravy Mix, Rich Onion Gravy Mix, Savoury Gravy Mix (for pork and chicken).

Heinz
Pickled Onions, Silverskin Onions.

Homepride
Cook-in-Sauce, White Wine with Cream Sauce.

Kashmir Brand
Madras Curry Paste, Vindaloo Curry Sauce, Mixed Pickle, Mango Pickle.

La Favorite
Dijon Mustard, French Mustard à L'Estragon, Sauce Vinaigrette, Surfine Capers, Tewkesbury Mustard.

Norco
Little Chop Piccalilli, Mixed Pickle, Piccalilli, Pickled Onions, Pickled Red Cabbage.

Patak
Diwali Curry Concentrates, mild, medium and hot, Vindaloo Paste, Lime Pickle Mild.

Rowats
Cocktail Gherkins, Cocktail Onions, Mixed Pickles, Pickled Cauliflower, Pickled Onions, Pickled Red Cabbage, Pickled Silverskin Onions, Stuffed Olives.

Sainsbury
Pickled Red Cabbage, Silverskin Onions.

Sharwood
Hot Curry Paste, Korma Mild Curry Sauce, Lemon Slices in Lemon Juice, Madras Hot Curry Sauce, Mild Curry Paste, Rogan Josh Medium Hot Curry Sauce, Sherry Olives in Brine, Stuffed Olives in Brine, Vindaloo Extra Hot Curry Sauce, White Wine Cooking Sauce, Worcestershire Sauce.

Shiva
Beef Curry (Medium Hot), Lime Pickle, Medium Curry Paste, Mixed Pickles, Vindaloo Curry Sauce.

Chocolate *See Chapter 12*

Boots
Diabetic Chocolate Father Christmases, Easter eggs, Easter rabbits.

Australian products

Beverages

Campbells
V8 Vegetable Juice.

Farmland
Apple Juice.

Nestlés
Malted Milk

Sanitarium
Caffex, Biscuits.

Nabisco
Premium, Wheat Toast.

Breakfast cereals

Cerebos
Breakfast D'Light

Kellogg's
Puffed Wheat, Ready-Wheats.

Sanitarium
Granola, Granose, Puffed Rice, Puffed Wheat, Wheat Germ.

Ice cream

Peters
Carbohydrate Modified Ice Cream.

Mustard, mayonnaise and savoury sauces

Cerebos
Gravox Chicken Sauce Mix, Gravy Powder, Supreme Gravy Powder, Gravox Parsley Sauce Mix

Keen's
Curry Powder – Standard, Madras, Vindaloo.

Masterfoods
English Hot Mustard, German Mustard.

Sanitarium
Gravy Quick.

Spreads

Farmland
Tomato Paste.

John West
Crab Spread.

Leggo's
Tomato Paste.

Kraft
Cheddar Cheese Spread, Vegemite – yeast extract.

Pecks
Anchovette, Salmon and Shrimp, Anchovy, Crab, Chicken, Salmon and Lobster.

Sanitarium
Peanut Butter

Weight Watchers
Jam (all flavours)

Appendix 2
Ingredients used in recipes

Bread

Dairy products
butter
cheese
   mature hard
   Edam
   skimmed milk soft
cream
margarine
natural low fat yoghurt
skimmed milk

Dried fruits
apricots
'industrial' apricots (cheaper, as tasty but need soaking in cold water for 30 minutes before using)
bananas
currants
dates
figs
prunes
raisins
sultanas

Dried goods
agar agar
baking powder
barley flakes
bran
cocoa powder
coffee
custard powder
desiccated coconut
gelatine
haricot beans
kibbled wheat
lentils
millet flakes
nuts
   almonds
   cashews
   hazels
   peanuts
   pecans
   walnuts
   creamed coconut
oatmeal
peas
porridge oats
rolled oats
rye flakes
semolina (wholewheat)
sesame seeds
short grain rice
shredded suet
Shredded Wheat
Suenut
sunflower seeds
wheat flakes
wheat germ
whole wheat
yeast

Eggs

Flours
buckwheat
cornflour
maize meal (or Indian corn)
plain
potato
rice
rye
self raising (81%)
soya
strong white
wholewheat

Fresh fruits
apples – cooking and eating
apricots
bananas – green and ripe
blackberries
blackcurrants
damsons
gooseberries
grapes
grapefruit
lemons
melon
oranges
peaches
pears
plums
raspberries
redcurrants
satsumas
strawberries
tangerines

Herbs, Spices and Flavourings
allspice
basil – fresh or dried
bay leaves
cardamom powder
cayenne pepper
celery seeds
chilli powder
cinammon – ground and stick
cloves – whole and ground
coriander – ground
cumin – ground
garlic
ginger – ground
lovage – seeds and leaves
mint
mustard – ground, made up and seeds
nutmeg – ground
oregano
paprika
parsley
pepper – ground
peppercorns
pickling spice
rose petals
saffron
salt
tarragon
turmeric
vanilla – essence and pods

Liquids
anchovy essence
brandy
evaporated milk
glycerin

juices
  unsweetened apple
  grape
  grapefruit
  orange
  pineapple
lemon juice
meat stock
mushroom ketchup
oil – vegetable and sunflower
orange flower water
rose water
sherry
tomato purée
vegetable stock
vinegar – wine, spirit, malt
  and cider
vodka
white rum

Meat & Fish
chicken
chicken livers
liver – lamb or calf
mince
white fish
smoked fish

Vegetables
asparagus
beetroot
broccoli
broad beans
cabbage
celery
cauliflower
carrots
courgettes
cucumber
kohl rabi
leeks
marrow
onions
parsnips
peas
peppers – red and green
potatoes
pumpkin
spring greens
sprouts
squash
swedes
tomatoes – fresh, canned and
  green
turnips
watercress

This list is provided as an *aide-memoire* since some of the ingredients are not readily available in supermarkets.

References

Preface

1 Medawar, P.B. *The Hope of Progress*. p. 99. 1972. Methuen.

Chapter 1

1 *Adult Dental Health in England and Wales in 1968*. HMSO. London.
2 *Adult Dental Health in Scotland 1972*. HMSO. London. 1974.
3 *Children's Dental Health in England and Wales 1973*. HMSO. London. 1975.
4 Kimmelmann, B.B. and Tassman, G.C. *J. Dent. Child*. **27**, p. 60. 1960.
5 Starkey, P. *J. Dent. Child*. **18**, p. 42. 1961.
6 McClure, D.B. *J. Dent. Child*. **33**, p. 205. 1966.
7 Sangnes, G. et al. *J. Dent. Child*. **34**, p. 94. 1972.
8 Frandsen, A.M. et al. *Scand. J. Dent. Res*. **78**, p. 459. 1970.
9 John Besford. *Good Mouthkeeping*. O.U.P. 1984.
10 Anderson, Jack D. *Personal communication*.
11 Sheiham, A. *Lancet* Aug 27 1977. p. 442.
12 *Children's Dental Health in England and Wales 1973*. HMSO. London. 1975.
13 Axelsson, P. and Lindhe, J. *J. Clin. Perio.*, **5**, pp. 133-151. 1978.
14 Holloway, P.J. *Int. Dent. J.*, **25**, pp. 26-30.
15 *Sugar and Dental Caries, a Policy Statement*. British Association for the Study of Community Dentistry. 1982.

Chapter 2

1 Moore, W.J. and Corbett, M.E. *The Distribution of Dental Caries in Ancient British Populations. 1. Anglo-Saxon Period* in *Caries Res*. **5**, pp. 151-168. 1971.

2 Ibid. II. *Iron Age, Romano-British and Mediaeval Periods* in *Caries Res.* **7**, pp. 139-153. 1973.

3 Ibid. III. *The 17th Century* in *Caries Res.* **9**, pp. 163-175. 1975.

4 James, P.M.C. and Miller, W.A. *Dental Conditions in a Group of Mediaeval English Children* in *Brit. Dent. J.* **128** (8).

5 Rush, M.A. *The Teeth of Anne Mowbray* in *Brit. Dent. J.* **119** (8).

Chapter 3

1 Stephan, R.M. *J. Amer. Dent. Ass.* **27**, p. 718.

2 Gustafsson, B.E. et al. *Vipeholm caries study* in *Acta Odont. Scand.* **11**, p. 232.

3 Stenton Doris Mary. *English Society in the Early Middle Ages*, Penguin. 1977.

4 Deerr Noel. *The History of Sugar*. Chapman and Hall. 1949.

5 Rush, M.A. *The Teeth of Anne Mowbray* in *Brit. Dent. J.* **119** (8).

6 Paul Hentzner. *A Journey into England, 1598* (trans. Horace Walpole, 1757).

7 Deerr Noel, Ibid. p. 284.

9 Orwell George, *Animal Farm*. Penguin. 1966.

10 Fairrie Geoffrey, *Sugar*. Fairrie & Co. Ltd. 1925.

11 Hardinge, M.G., Swarner, J.B. and Crookes, H., *Carbohydrates in Food* in *J. Am. Dietet. A.*, **46**, pp. 197–204.

12 Hartley Dorothy, *Food in England*. MacDonald. 1954. p. 351.

13 Drummond Sir Jack, and, Wilbraham Anne. *The Englishman's Food*. Cape. 1957. p. 465.

14 Jenkins, G.N. *Natural Protective Factors of Foods*, in *Symp.* **3**, pp. 67–74.

15 Hartley Dorothy, Ibid.

16 Geddes, D.A.M. et al. *Brit. Dent. J.* 1977. **142,** p. 317.

17 Jenkins, G.N. *The Physiology and Biochemistry of the Mouth*. 1978. p. 301.

18 Ibid. p. 457.

19 *Sugar and Dental Caries: a Policy Statement.* British Association for the Study of Community Dentistry. 1982.

20 Rugg-Gunn A.J. et al. *Brit. Dent J.* 1975. **139,** p. 351.

21 Geddes, D.A.M. et al. Ibid.

22 Jenkins, G.N. *Indent.*, Vol. 1, No. 1.
23 Ibid.

Chapter 4

1 Lehner T. *New Scientist*. p. 217. 27 April 1978.
2 Waller, H. *Arch. Dis. Child.* **21.** 1946.
3 Sedgwick, J.P. *Amer. J. Dis. Child.* **21**, p. 455. 1921.
4 DHSS. *Present-Day Practice in Infant Feeding. pp3–4.* HMSO.
5 MacKeith, R.R., Wood, C. *Infant Feeding and Feeding Difficulties*. Churchill Livingstone. p. 68.
6 Brooke, O.G. *Nutrition and Food Science.* **44.** July 1976 and *Personal Communication* April 1982.
7 DHSS. Ibid. p. 18.
8 Crandon, J.H. et al. *New Eng. J. Med.* **223,** p. 10.
9 *Sugar and Dental Caries: a Policy Statement.* British Association for the Study of Community Dentistry. 1982.
10 King, J.M. *Community Dent. Oral Epidemiol.* 1978, **6,** pp. 47–52. 1982.
11 MacKeith, Dr. S. Personal Communication to the author.

Chapter 5

1 Winter, G.B. et al. *Arch. Dis. Child.* 1966. **41,** p. 207.
2 Goose, D.H. and Gittus, E. *Public Health.* 1968. **82,** p. 2.
3 Gardner, E.E. et al. *J. Dent. Child.* 1977. **45,** p. 3.
4 MacKeith, R. and Wood C. *Infant Feeding and Feeding Difficulties*. Churchill Livingstone. 1977.

Chapter 6

1 MacKeith R. and Wood C. *Infant Feeding and Feeding Difficulties*. Churchill Livingstone. 1977.

Chapter 7

Suggested Reading
Septima. *Something to Do*. Penguin.
Starter Activities, Macdonald Educational. (Except "Cakes and Biscuits".

Something to Make. Young Puffin Original.
The Pre-School Child. Ward Lock.
Betts, Jan. *Knock at the Door*. Ward Lock Educational.
Craft, Ruth. *Play School Play Ideas*. BBC Publications.
You and Me. BBC Publications.

Chapter 8

1 David, Elizabeth. *English Bread and Yeast Cookery*. Allen Lane. Penguin Books Ltd. 1977.
2 Leach, Margaret. *Freezer Facts*. Forbes Publications Ltd. 1975.

Chapter 9

1 Milk Marketing Board. *What Are Children Eating These Days?* 1982.
2 *Confectionery in Perspective*, from the Cocoa, Chocolate and Confectionery Alliance, 11 Green Street, London W1Y 3RF.
3 Southgate, D.A.T. et al. *The Free Sugar Content of Some Foods*. Mimeo. 27 April 1978.
4 Macdonald, I. et. al. *Br. J. Nutr.*, **24**, p. 413. 1970.
5 Drummond, J.C. *The Englishman's Food*. Cape. p. 404. 1957.
6 Weaver, R. *Brit. Dent. J.* **88**, 9, p. 231.
7 Sheiham A. *J. Clin. Perio.* **6**, 7, p. 7.
8 Moore, W.J. and Corbett. E. *Caries Res.* **7**, pp. 139-153.
9 White, Florence. *Good Things in England*. Futura. pp. 339, 340. 1974.

Chapter 13

1 Fortune, R.F. *Sorcerers of Dobu*. Routledge and Kegan Paul. 1932.
2 Davies, D.M. *The Influence of Teeth, Diets and Habits on the Human Face*. p. 96. 1972. Heinemann.
3 Ibid. p. 98.
4 *The Scientific Basis of Dental Health Education*. The Health Education Council. 1978.
5 Kieser. J. B. and Wade, A.B. *J. Clin. Perio.* **3**, p. 200. 1976.

Acknowledgements

Chapter 1
Laurie Lee, from *Selected Poems*, reproduced by permission of Andre Deutsch. Originally published by John Lehman in 1947. Spike Milligan, *Silly Verse for Kids*, reproduced by permission of the author.

Chapter 3
Noel Deerr, *History of Sugar*, reproduced by permission of Chapman and Hall. George Orwell, *Animal Farm*, reproduced by permission of the estate of the late Sonia Brownell Orwell and Martin Secker and Warburg. Dorothy Hartley, *Food in England*, reproduced by permission of Mac-Donald.

Chapter 5
John Betjeman, *Collected Poems*, reproduced by permission of John Murray (Publishers) Ltd.

Chapter 6
MacKeith and Wood, *Infant Feeding Difficulties*, reproduced by permission of Churchill Livingstone.

Chapter 7
Norman Nicholson *The Pot Geranium* and Michael Roberts *Collected Poems*, reproduced by kind permission of Faber and Faber Ltd.

Chapter 8
A. P. Herbert, quote reproduced by permission of Lady Gwendolen Herbert. John Gunther *Inside Russia Today*, reproduced by permission of Hamish Hamilton Ltd.

Chapter 9
Henrik Ibsen, *A Doll's House*, reproduced by permission of Penguin Books Ltd.

Chapter 11
Magnus Pyke, *Food and Society*, reproduced by permission of John Murray (Publishers) Ltd.

Chapter 12

Roy Fuller, *Collected Poems (1962)*, reproduced by permission of Andre Deutsch Ltd.

Chapter 13

Robert Baden-Powell, reproduced by permission of The Hamlyn Publishing Group Ltd from *Scouting for Boys*, originally published by C. Arthur Pearson Ltd.

General index

*see Appendix 1 for
  recommended products

Index to recipes

Fiction

The Chains of Fate	Pamela Belle	£2.95p
Options	Freda Bright	£1.50p
The Thirty-nine Steps	John Buchan	£1.50p
Secret of Blackoaks	Ashley Carter	£1.50p
Hercule Poirot's Christmas	Agatha Christie	£1.50p
Dupe	Liza Cody	£1.25p
Lovers and Gamblers	Jackie Collins	£2.50p
Sphinx	Robin Cook	£1.25p
My Cousin Rachel	Daphne du Maurier	£1.95p
Flashman and the Redskins	George Macdonald Fraser	£1.95p
The Moneychangers	Arthur Hailey	£2.50p
Secrets	Unity Hall	£1.75p
Black Sheep	Georgette Heyer	£1.75p
The Eagle Has Landed	Jack Higgins	£1.95p
Sins of the Fathers	Susan Howatch	£3.50p
Smiley's People	John le Carré	£1.95p
To Kill a Mockingbird	Harper Lee	£1.95p
Ghosts	Ed McBain	£1.75p
The Silent People	Walter Macken	£1.95p
Gone with the Wind	Margaret Mitchell	£3.50p
Blood Oath	David Morrell	£1.75p
The Night of Morningstar	Peter O'Donnell	£1.75p
Wilt	Tom Sharpe	£1.75p
Rage of Angels	Sidney Sheldon	£1.95p
The Unborn	David Shobin	£1.50p
A Town Like Alice	Nevile Shute	£1.75p
Gorky Park	Martin Cruz Smith	£1.95p
A Falcon Flies	Wilbur Smith	£2.50p
The Grapes of Wrath	John Steinbeck	£2.50p
The Deep Well at Noon	Jessica Stirling	£2.50p
The Ironmaster	Jean Stubbs	£1.75p
The Music Makers	E. V. Thompson	£1.95p

Non-fiction

The First Christian	Karen Armstrong	£2.50p
Pregnancy	Gordon Bourne	£3.50p
The Law is an Ass	Gyles Brandreth	£1.75p
The 35mm Photographer's Handbook	Julian Calder and John Garrett	£5.95p
London at its Best	Hunter Davies	£2.95p
Back from the Brink	Michael Edwardes	£2.95p

☐	**Travellers' Britain**	⎱ Arthur Eperon	£2.95p
☐	**Travellers' Italy**	⎰	£2.95p
☐	**The Complete Calorie Counter**	Eileen Fowler	80p
☐	**The Diary of Anne Frank**	Anne Frank	£1.75p
☐	**And the Walls Came Tumbling Down**	Jack Fishman	£1.95p
☐	**Linda Goodman's Sun Signs**	Linda Goodman	£2.50p
☐	**Scott and Amundsen**	Roland Huntford	£3.95p
☐	**Victoria RI**	Elizabeth Longford	£4.95p
☐	**Symptoms**	Sigmund Stephen Miller	£2.50p
☐	**Book of Worries**	Robert Morley	£1.50p
☐	**Airport International**	Brian Moynahan	£1.75p
☐	**Pan Book of Card Games**	Hubert Phillips	£1.95p
☐	**Keep Taking the Tabloids**	Fritz Spiegl	£1.75p
☐	**An Unfinished History of the World**	Hugh Thomas	£3.95p
☐	**The Baby and Child Book**	Penny and Andrew Stanway	£4.95p
☐	**The Third Wave**	Alvin Toffler	£2.95p
☐	**Pauper's Paris**	Miles Turner	£2.50p
☐	**The Psychic Detectives**	Colin Wilson	£2.50p
☐	**The Flier's Handbook**		£5.95p

All these books are available at your local bookshop or newsagent, or
can be ordered direct from the publisher. Indicate the number of copies
required and fill in the form below   11

....

Name_____
(Block letters please)

Address_____

_____

Send to CS Department, Pan Books Ltd, PO Box 40, Basingstoke, Hants
Please enclose remittance to the value of the cover price plus:
35p for the first book plus 15p per copy for each additional book ordered
to a maximum charge of £1.25 to cover postage and packing
Applicable only in the UK

While every effort is made to keep prices low, it is sometimes
necessary to increase prices at short notice. Pan Books reserve
the right to show on covers and charge new retail prices which
may differ from those advertised in the text or elsewhere

Cook books

☐	**The Infra-Red Cook Book** Kathy Barnes	£1.50p
☐	**Mrs Beeton's Cookery For All** Mrs Beeton	£3.95p
☐	**The Microwave Cook Book** Carol Bowen	£1.95p
☐	**Pressure Cooking Day by Day** Kathleen Broughton	£2.50p
☐	**Middle Eastern Cookery** A. der Haroutunian	£2.95p
☐	**Vegetarian Cookbook** Gail Duff	£2.95p
☐	**Crockery Pot Cooking** Theodora Fitzgibbon	£1.50p
☐	**The Book of Herbs** Dorothy Hall	£1.95p
☐	**The Best of Cordon Bleu** Rosemary Hume and Muriel Downes	£1.95p
☐	**Diet for Life** Mary Laver and Margaret Smith	£1.95p
☐	**Herbs for Health and Cookery** Claire Loewenfeld and Philippa Back	£2.50p
☐	**The Preserving Book** Caroline Mackinlay	£4.50p
☐	**The Book of Pies** Elisabeth Orsini	£1.95p
☐	**Learning to Cook** Marguerite Patten	£2.50p
☐	**Wild Food** Roger Phillips	£5.95p
☐	**Complete International Jewish Cookbook** Evelyn Rose	£2.95p
☐	**Caribbean Cookbook** Rita Springer	£1.95p
☐	**The Times Cookery Book** } Katie Stewart	£3.50p
☐	**Shortcut Cookery**	£1.95p
☐	**Freezer Cookbook** Marika Hanbury Tenison	£1.95p
☐	**The Pan Picnic Guide** Karen Wallace	£1.95p

All these books are available at your local bookshop or newsagent, or can be ordered direct from the publisher. Indicate the number of copies required and fill in the form below   11

....

Name_____
(Block letters please)

Address_____

_____

Send to CS Department, Pan Books Ltd, PO Box 40, Basingstoke, Hants
Please enclose remittance to the value of the cover price plus:
35p for the first book plus 15p per copy for each additional book ordered
to a maximum charge of £1.25 to cover postage and packing
Applicable only in the UK

While every effort is made to keep prices low, it is sometimes
necessary to increase prices at short notice. Pan Books reserve
the right to show on covers and charge new retail prices which
may differ from those advertised in the text or elsewhere